Plant Biotechnology: Principles and Applications

Plant Biotechnology: Principles and Applications

Nathan Mitchell

SYRAWOOD
PUBLISHING HOUSE

New York

Published by Syrawood Publishing House,
750 Third Avenue, 9th Floor,
New York, NY 10017, USA
www.syrawoodpublishinghouse.com

Plant Biotechnology: Principles and Applications
Nathan Mitchell

International Standard Book Number: 978-1-64740-008-8 (Hardback)

Cataloging-in-Publication Data

Plant biotechnology : principles and applications / Nathan Mitchell.
    p. cm.
Includes bibliographical references and index.
ISBN 978-1-64740-008-8
1. Plant biotechnology. 2. Agricultural biotechnology. 3. Plants.
I. Mitchell, Nathan.
SB106.B56 P53 2020
630--dc23

TABLE OF CONTENTS

Permissions

Index

PREFACE

Plant biotechnology is a field of agricultural science that makes use of scientific tools and techniques for the purpose of modifying plants. Some of the techniques and tools used within this field are genetic engineering, molecular markers, vaccines, molecular diagnostics and tissue culture. One of its major sub-domains is crop biotechnology where a desired trait from one species of plant is added to an entirely different species. These desired characteristics include flavor, growth rate and resistance to diseases and pests. There are diverse modification techniques which are used in plant biotechnology such as mutagenesis, polyploidy, protoplast fusion, transgenics and genome editing. This book elucidates the concepts and innovative models around prospective developments with respect to plant biotechnology. It aims to shed light on some of the unexplored aspects of this field. Coherent flow of topics, student-friendly language and extensive use of examples make this book an invaluable source of knowledge.

A detailed account of the significant topics covered in this book is provided below:

Chapter 1- Plant biotechnology is involved in the processes of breeding and genetically engineering crops. This is an introductory chapter which will introduce briefly all the significant aspects of plant biotechnology such as its role in the production of genetically engineered crops.

Chapter 2- The collection of techniques used to grow or maintain plant cells, tissues and organs under sterile conditions on a nutrient culture medium of known composition is known as plant tissue culture. Some of the different types of plant tissue culture are seed culture, embryo culture and callus culture. This chapter discusses in detail these concepts and processes related to plant tissue culture.

Chapter 3- The science of changing the traits of plants in order to produce desired characteristics is known as plant breeding. It is primarily used to improve the quality of nutrition for humans and animals. It is also involved in making the plants disease and drought resistant. The following chapter elucidates these varied processes and mechanisms associated with the breeding of plants.

Chapter 4- A fragment of DNA which is associated with a certain location within a genome is known as a molecular marker. They are used in biotechnology and molecular biology in order to make the breeding process much more efficient and speedy. This chapter has been carefully written to provide an easy understanding of the various applications of molecular markers for the purpose of breeding.

Chapter 5- The method used to insert DNA from another plant, into the genome of a plant of interest is known as plant transformation. Plant transformation vectors are the plasmids that are created to facilitate the generation of transgenic plants. The chapter closely examines these key concepts of plant transformation technology to provide an extensive understanding of the subject.

Chapter 6- Plant biotechnology is used in a variety of different areas such as improving the nutritional quality of different food crops, improving the flavor and texture of fruits and vegetables, and in the process of brewing and malting. It is also involved in the production of industrial enzymes. All these diverse concepts and processes related to plant biotechnology have been carefully analyzed in this chapter.

It gives me an immense pleasure to thank our entire team for their efforts. Finally in the end, I would like to thank my family and colleagues who have been a great source of inspiration and support.

Nathan Mitchell

Chapter 1

Plant Biotechnology

Plant biotechnology is involved in the processes of breeding and genetically engineering crops. This is an introductory chapter which will introduce briefly all the significant aspects of plant biotechnology such as its role in the production of genetically engineered crops.

Agricultural Biotechnology

Agricultural biotechnology is the use of different scientific techniques to modify plants and animals. The undesirable characteristics like susceptibility to diseases and low productivity are bred out. If there is a particular trait that the plant or animal can benefit from, it can be bred in by using a gene that contains the characteristic.

Biotechnology has especially been beneficial in improving agricultural productivity and increasing the resistance of plants to diseases. Scientists do this by studying the DNA. They first identify the gene that would be beneficial to the plant or animal then work with the characteristics conferred in a precise and exact manner to achieve the desired outcome.

Advantages of Agricultural Biotechnology

Biotechnology has been beneficial in many ways. First, stabilized plants that have higher yields have been produced successfully. The resistance of these plants to pests, diseases and abiotic factors such as rainfall has played a major role in increasing the yields.

Animal feeds are being improved by biotechnology to increase their nutrient intake and reduce environmental wastes.

Another advantage of biotechnology is that it has led to the development of better vaccines that don't necessarily have to be stored in very cold temperatures. Penicillin, one of the most important components of antibiotics was produced through biotechnology.

How Agricultural Biotechnology is used

Genetic Engineering

Genetic engineering, also referred to as genetic improvement or modification, is the movement of a gene from one organism to another. This process allows for the transfer of a useful characteristic into an organism by inserting it with a gene containing the particular trait. In crops, genetic engineering has been used to increase productivity and resistance to weeds and harsh weather conditions.

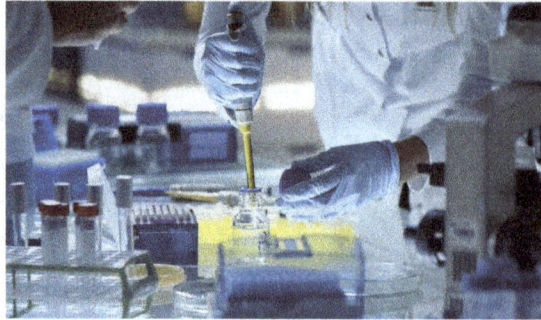

Molecular Markers

Breeding was previously concerned with the removal and insertion of desirable physical traits, an example being the aggression of Bulldogs. By studying the DNA, scientists were able to find molecular markers that showed traits that were not visible. Using molecular markers, breeding has been made more precise and accurate, and this has countered the undesirable characteristics that may have appeared in future generations.

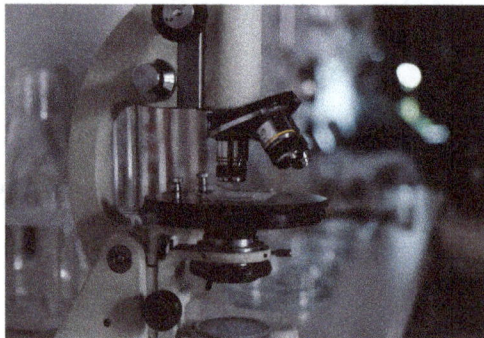

Vaccines

Biotechnology is used for making vaccines for both animals and human beings. These vaccines are better than the traditional ones because they are cheaper, safer, and can survive warmer tropical temperatures. Vaccines to prevent new infections have also been developed using biotechnology.

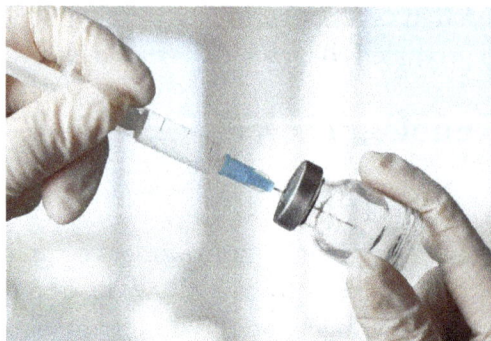

Genomics

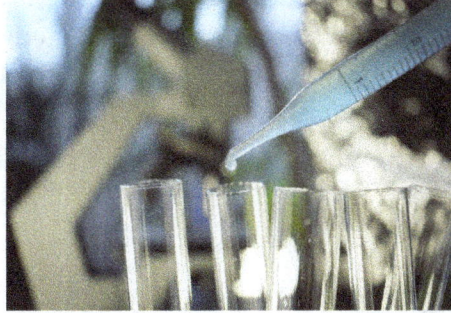

A genome is an entire set of chromosomes found in the DNA, and through the study of genomes and genetic mechanisms, breakthroughs have been made in biotechnology. Through genomics, the structure, function, location, and impact of a particular gene and genome are identified. This makes it easy to determine the characteristics that will be transferred to another organism and the exact results of the transfer of the gene.

Tissue Culture

This technique is used to produce a plant that is free from undesirable characteristics, which are mostly, diseases. A disease-free plant part is used to generate types that are disease-free. The different types of plants in which tissue culture works include bananas, avocados, mangoes, coffee, and papaya, among others.

Biotechnology in Agriculture and Food Production

Biotechnology has positively influenced the economy and social life of most developing countries. The increased food production by biotech plants means that more people can now enjoy food security while spending less on pesticides. This has, subsequently, led to increased standards of living.

Biotechnology in agriculture, both for plants and for animals is a reality. It offers a tool through which these organisms can be understood, and their genetic resource management improved. Plants and animals have, thus been able to increase their productivity and get better resistance to diseases through the study of their genes and the manipulation of their characteristics. There are different ways through which biotechnology is done, and they include genetic engineering, vaccination, molecular marker, tissue culture, and genomics. Technology is evolving and so does the world of science, and who knows, there might be another fascinating breakthrough in genetics and biotechnology.

Agricultural Biotechnology Products

Vaccines

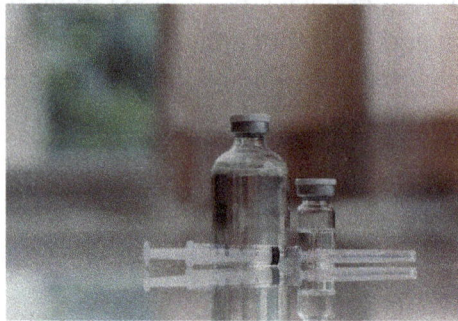

New vaccines employing biotechnology innovations are changing the processes of preventing illnesses, particularly in developing countries. Genetically modified crops have had a significant contribution in the development of vaccines. Foods such as fruits, grains, and vegetables are engineered to carry antigenic proteins which are extracted from pathogens. When injected into the body, these antigens trigger an immune response and boost the resistance of the body against the pathogens.

An example is the anti-lymphoma vaccine that's obtained from tobacco. Tobacco plants are engineered to carry RNA from malignant B-cells. The extracted protein is injected into the body, an immune response is triggered which destroys the cancerous cells.

Plant and Animal Reproduction

The use of traditional techniques such as cross-pollination, grafting, and cross-breeding to enhance the behavioral patterns of plants and animals is time-consuming. Agro-biotech has made it possible to enhance plant and animal traits on a molecular level through over-expression or gene removal, or the introduction of foreign genes.

Artificial insemination, embryo transfer, and other associated technologies are used in managing the reproductive functions of an animal and influencing the traits of the resultant offspring. These improvements have increased agricultural productivity in developing countries and enhanced their capabilities to sustain the growing population.

Antibiotics

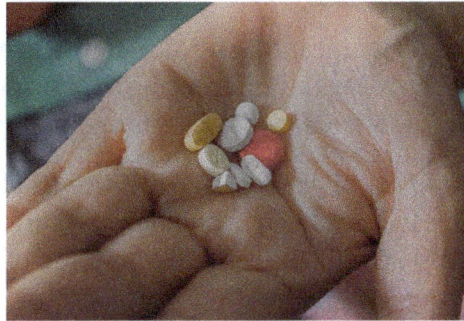

Agricultural biotechnology is applied in the production of antibiotics for both humans and animals. Animal antibiotics produced through this technology are low cost but equally as efficient as traditionally manufactured antibiotics. Since these antibiotics are obtained from plants, a large quantity of the product can be obtained at a time. Additionally, there is ease of purification and the risk of contamination is minimized as compared to the use of mammalian cells and culture media in antibiotics production.

Nutritional Supplements

In a bid to promote better human health globallys, scientists have come up with ways to create genetically modified foods with nutrients that can help fight disease and starvation. A great example of such foods is the golden rice which contains beta carotene, a major source of Vitamin A in the body.

The name of the rice comes from the color of the transgenic grain made from three genes: two from daffodils and one from bacterium. The genes are cloned to make the rice "golden." People who eat this rice supplement their diet with the vitamin and other nutrients that they may not be getting from other foods.

Pesticide-resistant Crops

In the past, farmers have incurred significant losses due to the use of pesticides that affect both crops and weeds. Biotechnology has led to the engineering of plants that are resistant to pesticides. This allows farmers to selectively kill weeds without harming their crop.

The tech was first introduced in genetically modified soy beans, making them resistant to the herbicide glyphosate. The herbicide can be applied in copious amounts to eliminate other plants on a field other than the Roundup-Ready plants. Selective elimination of weeds saves farmers' valuable time as compared to traditional methods of weeding.

Pest-resistant Crops

For many years, a microbe known as Bacillus thuringiensis (Bt) has been used to dust crops by producing toxic proteins against pests. One of such toxic proteins used for dusting crops is the European corn borer. Scientists have come up with a way to eliminate the use of Bt by introducing pest resistant crops. These are known as Bt crops as the gene that's introduced in the crop was originally identified in Bacillus thuringiensis. Examples if pest resistant crops today are Bt maize, potato, and corn. This toxic protein is only harmful to pests, but is safe for humans. It has saved farmers from dealing with expensive pest infestations in crops.

Flowers

Agricultural biotechnology is not just about developing drugs and genetically modified foods and crops – it has some aesthetic applications as well. Scientists are using gene recognition and transfer techniques to improve the color, size, smell, and other properties of flowers. The technology has also been used to improve other ornamental plants such as shrubs and trees. Some of the techniques applied are similar to those used on crops. For instance, tropical plants' color confrontation can be enhanced to make it possible for the tree to thrive in gardens in the northern regions.

Bio-fuels

The agricultural industry plays a major role in the production of bio-fuels to the extent that feed-stock is used for fermentation and purification of bio-oil, bio-ethanol, and bio-diesel. Genetic engineering and enzyme optimization techniques are used to produce good-quality feed stocks for more efficient conversions and higher BTU outputs of the resultant bio-fuels. High-energy and high-yielding feed stocks can reduce the relative costs of harvesting and transportation. The result is high-quality bio-fuel products.

A Biotic Strain Confrontation

A very small proportion of the earth, approximately 20 percent, is arable land. However, scientists have come up with ways to modify crops that can endure conditions such as salinity, cold, and drought. For instance, the detection of genes in plants that are tasked with the uptake of sodium has led to the introduction of plants that can thrive in high-salinity environments.

A technique known as up- or down-regulation of record is used to influence drought-resistance in plants. These technologies have increased food production as plants are able to adapt to hostile climates and non-arable lands.

Manufacture of Power Fibers

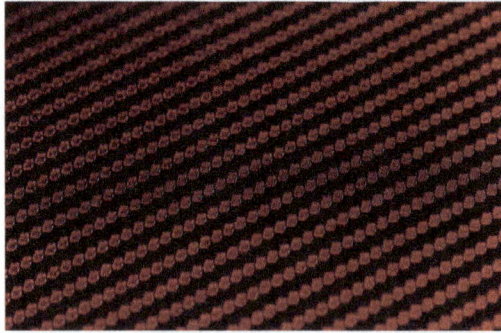

The strongest fiber known to man today is spider silk. It is stronger than kevlar, which is used to make bullet-proof vests and has a higher tensile strength than strength.

For a while, the technology seemed like it would solve the problem. However, it was shelved when the scientists couldn't come up with a way to spin the protein into fibers. Although the technology has been put on hold, it is bound to appear again in the future because of high demand for similar products.

Plant Biotechnology

There have always been efforts to improve upon the plant characteristics by controlled sexual breeding within the same species.

Plant breeding is based on combining the hereditary information (which regulates the desirable phenotypes) of both parents and passing it on to the offspring. Conventional breeding normally takes a long time to improve upon a trait. Further, it does not permit the mixing of incompatible genes, and cross breeding of unrelated genes yields infertile offspring.

But modem biotechnology has scaled these problems. The new tools of genetic engineering can transfer the desired gene or a set of genes enduring a specific function. This technology makes it possible to break through the species barrier and shuffle information between completely unrelated species in a controlled manner. Such technology is referred to as precision breeding.

Principles and Applications

Over the last one decade, there has been remarkable progress in research and development in this field. Significant work has been done on the isolation, manipulation and subsequent growth of naked plant cells (protoplasts) and cells in tissue culture. The second field, the recombinant (rDNA) technology or genetic engineering has emerged out of the work initially carried out on micro-organisms.

Now to achieve such results using the existing technologies one, need to have the vector with the gene of interest ready before it could be transformed further into the plant. The three major techniques have been used to transfer genes of interest in plants. These include Agrobacterium mediated approach, protoplast based approach and biolytic or gene gun approach.

Genetically Engineered Crops

Genetic Engineering is the introduction of a specific gene into the DNA of a plant to obtain a desired trait. The gene introduced may come not only from another plant species, but also from other organisms. While traditional plant breeding involves crossing related plants, biotechnology is a new tool that enhances the capability of breeders to be more precise.

What are the Goals of Genetic Engineering?

The goals of genetic engineering are the same as with traditional breeding. They may aim to improve crop performance in the field by conferring pest and disease resistance, herbicide resistance, or tolerance to environmental stresses (such as drought or flooding). They may also aim to develop products with enhanced value, such as improved post-harvest life, nutritional value, or other health benefits.

Insect Resistance

In the last few years, several crops have been genetically engineered to produce their own Bt proteins, making them resistant to specific groups of insects. "Bt" is short for Bacillus thuringiensis, a soil bacterium that contains a protein that is toxic to a narrow range of insects, but not harmful to animals or humans. Applications of Bt bacteria have been used to control insect pests for many years, before the advent of the current Bt crops made using biotechnology.

Varieties of Bt insect-resistant corn and cotton are now in commercial production. Other crops being investigated include cowpeas, sunflower, soybeans, tomatoes, tobacco, walnut, sugar cane, and rice.

Bt Corn field.

Herbicide Tolerance

Chemical herbicides are frequently used to control weeds. Weeds growing in the same field with crop plants can significantly reduce crop yields because the weeds compete for soil nutrients, water, and sunlight. Many farmers now control weeds by spraying herbicides directly onto the crop plants. Because these herbicides generally kill only a narrow spectrum of plants (if they didn't, they would kill the crop plants, too), farmers apply mixtures of multiple herbicides to control weeds after the crop has started to grow. It is often argued that such GE varieties reduce soil erosion, because they make adoption of soil-conserving practices such as "no-till" easier. Resistance to synthetic herbicides has been genetically engineered into corn, soybeans, cotton, canola, sugar beets, rice, and flax. Some of these varieties are commercialized in several countries. Research is ongoing on many other crops. One application of this technology is that herbicide could be coated on seed from an herbicide resistant variety (for example, maize) and while the maize would germinate and thrive, weeds and parasites such as Striga would be killed.

Virus Resistance

Many plants are susceptible to diseases caused by viruses, which are often spread by insects (such as aphids) from plant to plant across a field. The spread of viral diseases can be very difficult to control and crop damage can be severe. Insecticides are sometimes applied to control populations of transmitting insects, but often have little impact on the spread of the disease. Often the most effective methods against viral diseases are cultural controls (such as removing diseased plants) or plant varieties bred to be resistant (or tolerant) to the virus, but such strategies may not always be practical or available. Scientists have discovered new genetic engineering methods that provide resistance to viral disease where options were limited before.

Delayed Fruit Ripening

Delaying the ripening process in fruit is of interest to producers because it allows more time for shipment of fruit from the farmer's fields to the grocer's shelf, and increases the shelf life of the fruit for consumers. Fruit that is genetically engineered to delay ripening can be left to mature on the plant longer, will have longer shelf-life in shipping, and may last longer for consumers.

Foods with Improved Nutritional Value

Genetic modification can be used to produce crops that contain higher amounts of vitamins to improve their nutritional quality. Genetically altered "golden rice," for example, contains three transplanted genes that allow plants to produce beta-carotene, a compound that is converted to vitamin A within the human body. Vitamin A deficiency—the world's leading cause of blindness—affects as many as 250 million children. Biotechnology has also been used to alter the content of many oil crops, either to increase the amount of oil or to alter the types of oils they produce. Biotechnology could also be used to upgrade some plant proteins now considered incomplete or of low biological value because they lack one or more of the 'essential' amino acids. Examples include maize with improved protein balance and sweet potatoes with increased total protein content. Reducing toxicity of certain foods is also a goal of biotechnology. For example, reduction of the toxic cyanogens in cassava has been shown to be possible and could be produced in the future.

References

- What-is-agricultural-biotechnology: mixerdirect.com, Retrieved 21 April, 2019

- Products-of-agricultural-biotechnology: mixerdirect.com, Retrieved 15 January, 2019

- Plant-biotechnology, yourarticlelibrary.com, Retrieved 21 July, 2019meaning, principle-and-application-of-plant-biotechnology, biotechnology:

- Warp-briefs-eng-scr: absp2.cornell.edu, Retrieved 7 May, 2019

Chapter 2

Tissue Culture: Technologies and Applications

The collection of techniques used to grow or maintain plant cells, tissues and organs under sterile conditions on a nutrient culture medium of known composition is known as plant tissue culture. Some of the different types of plant tissue culture are seed culture, embryo culture and callus culture. This chapter discusses in detail these concepts and processes related to plant tissue culture.

Plant Tissue Culture

Plant tissue culture or micropropagation technology has made invaluable contribution to agriculture by enabling the production of disease free, quality planting material of commercial plants and fruit trees, throughout the year. It is a technique for in-vitro growth of plantlets from any part of the plant in a suitable nutrient medium under controlled aseptic conditions.

The success of agriculture development hinges on selection of desired types of plants and their multiplication. Traditionally, agriculture crops are multiplied by means of seeds (sexual propagation) or organs other than seeds (asexual or vegetative propagation). These organs are usually stems, roots or modified underground structures. Though multiplication by seeds is the cheapest method, it suffers form certain disadvantages. Plants raised from seeds may not repeat good performance of mother plants.

Many horticultural plants take a long time to produce seeds/fruits and many of them do not produce viable seeds or desired quality of seeds. Plants propagated vegetatively do not suffer from these disadvantages. However, vegetative propagation is rather a slow, time and space consuming process. Besides, it is usually infected with latent diseases. Some plants are also not amenable to vegetative method of propagation, for example, coconut, papaya, oil palm, clove etc.

Major Advantages of Tissue-culture

The main advantage of tissue culture technology lies in the production of high quality and uniform planting material that can be multiplied on a year-round basis under disease-free conditions anywhere irrespective of the season and weather. However, the technology is capital, labour and energy intensive. Although, labour is cheap in many developing countries, the resources of trained personnel and equipment are often not readily available. In addition, energy, particularly electricity, and clean water are costly. The energy requirements for tissue culture technology depend on day temperature, day-length and relative humidity, and they have to be controlled during the process of propagation. Individual plant species also differ in their growth requirements. The commercial advantages of tissue culture technology over its conventional counterpart are summarized below:

- Tissue culture could be a useful way for circumventing or eliminating disease, which can accrue in stock plants.

- Tissue Culture Plants (TCPs) may have increased branching and flowering, greater vigour and higher yield, mainly due to the possibility of elimination of diseases.

- The method may succeed to propagate plants where seeds or vegetative propagation is not possible or difficult or undesirable. As the capital investment on mother plants is reduced to almost zero, it may be easier to adapt to changing conditions. Additionally, a better programming of the production is possible, because of the greater plant uniformity and the availability in the mass at any time.

- Enables storage and maintenance of stock plants/germplasm.

Tissue Culture Technology

Tissue culture technology is based on the theory of totipotency i.e. the ability of a single cell to develop into whole plant. The major components of the technology include choice of explant (excised part of plant), growing of explant on a defined medium in glass vessel (in vitro), elimination and or prevention of diseases, providing appropriate cultural environment and transfer of plantlets from glass vessel to natural environment (hardening). All these constitute protocol for tissue culture. It varies from species to species and variety to variety within the same species. However, it can be standardized through trial and error and ultimately it should be repeatable and reliable.

Stages of Tissue Culture

Propagation by tissue culture is divided into five stages. A general account of these stages is outlined below.

Choice of Explant

Explants could be shoot tips (meristem), nodal buds, sections from internodes, leaves, roots, centres of bulbs, corms or rhizomes or other organs. The choice depends on the species to be multiplied and the method of shoot multiplication to be followed. Actively growing (shoot tips), juvenile (seedlings) or rejuvenated tissues (suckers) are preferred.

The commercial tissue culture laboratories commonly use tips of apical or lateral shoots, which contain meristems. Meristems are made up of actively dividing cells in an organized manner. They are about 0.1 mm in diameter and 0.25 - 0.30 mm. in length. However, explants should be chosen from typical, healthy, disease free, well-tested mother plants cultivated under conditions, which reduce contamination and promote growth of tissues to be cultured. If necessary explants may be subjected to virus testing and elimination. The selection of mother plants is very important for commercial success of tissue culture propagation.

Establishment of Germfree (Aseptic/Sterile) Culture

Excised part of plant is surface sterilized and transferred to sterile nutrient medium contained in glass vessel. On an average, about 25 cc nutrient media may be added per glass vessel. The cultures are maintained in growth rooms. If there is no infection and tissue isolated from mother plants survive in the artifical environment, initiation of new growth will take place after a week or so. Thus, germ-free culture is established.

Production of Shoots/Propagules

Once growth is initiated by induction of meristematic centres, buds develop into shoots by multiplication of cells. There are three types of multiplication systems for production of shoots.

Multiplication by Axillary Shoots

In this case shoots are produced from excised shoot tips or nodes. Commonly hormones (cytokinins) are used to induce multiple branching wherein, the rate of multiplication is low. Still it is preferred, because axillary shoots are likely to be genetically stable and the chances of production of types unlike mothers are less.

Multiplication by Adventitious Shoots

Explants such as sections of leaves, internodes or roots can produce directly adventitious shoots or other organs. This system has higher multiplication rate, but lesser genetic stability than axillary system.

Multiplication by Somatic Embryos (Embryoids)

Embryos are usually formed by the union of male and female reproductive cells (zygotic embryo), which ultimately can develop into a young plant. Embryo - like structures can also be produced from somatic cells. Somatic embryos are independent bipolar structures and are not attached to the tissues of origin. They can also develop to form young plants like zygotic embryos. Somatic embryos may be produced directly from explants such as sections of leaves, internodes or roots on solid culture medium.

The most common form of regeneration of plants occurs indirectly from callus. Callus is a mass of undifferentiated dividing cells often formed in tissues cultured in-vitro. Callus may give rise either to adventitious shoots, which develop into plantlets, or somatic embryos, which develop into seedlings. Callus is formed even naturally in response to wound.

Selecting proper tissue and culture medium can induce the formation of callus. This system has the highest multiplication rate and produce complete tiny plants. One gram of explants can produce one lakh somatic embryos. Dormancy can be induced in them or they can be transformed into synthetic seeds. However, callus is genetically unstable or plants arising from it may be unlike mother plants.

Such plants are known as off-types. They occur more frequently in callus culture and adventitious shoot culture as compared to axillary shoot culture. Off-types are undesirable in commercial propagation. Regeneration of shoots or intact plants by any one of the multiplication systems described above is influenced by many factors, such as composition of medium (specially concentration of growth regulators), type of tissue, genotype, ploidy level, etc.

Normally, multiplication cycle i.e., the period from incubation of plant parts on medium to formation of shoots varies from 3 to 6 weeks. However, the process is recycled many times by sub-culturing in order to obtain required multiplication rates. After completion of a cycle, shoots are cut separately and transferred to fresh medium. Cutting is done manually by using dissecting tools in laminar flow cabinets, where the air is clean to prevent any contamination. Once the shoots are placed on fresh medium, they are transferred back to the growth rooms. Thus, it may be possible to multiply the shoots 3 to 10 times per cycle of 3 to 6 weeks duration.

Preparation of Micro-cuttings for Establishment in the Natural Environment

Young axillary or adventitious shoots are finally separated from clusters (micro cutting) for initiation and development of roots. After separation, they are transferred individually to a medium containing rooting hormone (auxin) and continued to be maintained in the growth rooms until the roots are formed. It may also be possible to transfer the micro cuttings directly to soil or compost in humid green house for root formation. Somatic embryos may directly develop into seedlings.

Establishment in the Natural Environment

The most critical stage in propagation by tissue culture is the establishment of the plantlets in the soil. The steps involved are as under:

- Washing of media from plantlets,
- Transfer of plantlets to compost/soil in high humid green house,
- Gradual decrease in humidity from 100% to ambient levels over 3-4 weeks,
- And gradual increase in light intensity.

Plantlets during their growth in laboratory do not photo synthesize and their control of water balance is very weak. They use sugar contained in medium as source of energy. They exist like bacteria (heterotrophs). They need to be converted to more plant like existence (autotrophs) i.e., they should be in a position to utilize carbon-di-oxide from the air and solar energy for their food requirement. This acclimatization on the harsh real environment, outside artifical laboratory milieu takes place gradually.

Culture Environment

Environment conditions in the growth room, which influence cell multiplication, are light, day length and temperature. In tissue culture, light is required for synthesis of green pigment (chlorophyll) and development of organs. The range of light intensities appropriate for culture room varies from 1000 to 5000 lux. Requirement of day length would be in the range of 16-18 hours. Temperature requirement varies from 20 – 30 °C depending on species of plants. Tropical plants may require higher temperature than temperate plants.

Prevention of Contamination

Prevention of contamination in tissue culture is extremely important for commercial success of the unit. The entire production can go waste if the culture is contaminated. Sugar rich culture medium, excised plant tissue and culture environment are all conducive to the growth of pathogens. Therefore, it is essential that all operations be conducted in sterile or aseptic conditions. Various stages involved in prevention of contamination are outlined below:

- Mother plants should be grown under conditions which do not promote diseases.

- Explants should be free from disease. Meristem, used as an explant, is usually free from disease. Surface sterilization of explants in solutions of sodium or calcium hypochlorite is necessary. Heat or treatment with certain chemicals may eradicate latent viruses. All equipments and culture media are sterilised by autoclaving at 15-lb/sq. inch pressure at 121 °C for 15 – 20 minutes. Double distilled water should be used for washing explant and preparation of culture medium. UV lamps assist in sterilisation of laminar flow cabinets, hatches and instruments.

- Air handling units are employed for growth rooms and culture transfer rooms in order to avoid cross contamination between different areas of operation inside the clean area. The sterile condition is obtained in laminar airflow cabinets as they are provided with special type of international standard HEPA filters. These filters remove all the dust particles of above 0.3 micron present in the air.

Types of Plant Tissue Culture

The following points highlight the top six types of tissue culture. The types are: 1. Seed Culture 2. Embryo Culture 3. Callus Culture 4. Organ Culture 5. Protoplast Culture 6. Anther Culture.

Seed Culture

Seed culture is an important technique when explants are taken from in vitro-derived plants and in propagation of orchids. Sterilising procedures are needed for plant materials that are to be used directly, as explant source can cause damage to tissues and affect regeneration. In that case, culture of seeds to raise sterile seedling is the best method. Orchid cloning in vivo is a very slow process.

Thus seeds can be germinated in vitro and vegetatively propagated by meristem culture is then carried out on a large scale. Most orchids are sown in vitro because: orchid seeds are very small and contain very little or no food reserves. Their small size (1.0-2.0 mm long and 0.5-1.0 mm wide) makes it very likely that they can be lost if sown in vivo, and the limited food reserves also make survival in vivo unlikely.

The seed consists of a thickened testa, enclosing an embryo of about 100 cells. The embryo has a round or spherical form. Most orchid seeds are not differentiated: there are no cotyledons, roots and/or endosperm.

The cells of an embryo have a simple structure and are poorly differentiated:

- Sowing in vitro makes it possible to germinate immature orchid embryos, thus shortening the breeding cycle.

- Germination and development take place much quicker in vitro since there is a conditioned environment and no competition with fungi or bacteria.

Orchid seeds imbibe water via the testa and become swollen. After cell division begun, the embryo cracks out of the seed coat. A protocorm-like structure is formed from the clump of cells and on this a shoot meristem can be distinguished.

Protocorm has a morphological state that lies between an undifferentiated embryo a shoot. Protocorms obtained by seed germination have many close similarities with those produced from isolated shoot tips; the term protocorm like-bodies has introduced when cloning orchids by meristem culture.

The vegetative propagation of orchids follows culture of seeds, transformation of meristem into protocorm-like bodies, and the propagation of protocorms by cutting them into pieces and the development of these protocorms to rooted shoots. A large number of factors influence the germination and growth of orchids. The mineral requirement of orchids is generally not high and a salt poor medium of Knudson and Vacin and Went are good.

Some of the orchids require (Paphiopedilum ciliolare) for germination while others requires low irradiance. Sugar is extremely important as an energy source, especially for those that germinate in darkness. Regulators are usually not necessary for seed germination, and their addition often leads to unwanted effects like callus formation, adventitious shoot formation, etc.

Embryo Culture

Embryo culture is the sterile isolation and growth of an immature or mature embryo in vitro, with the goal of obtaining a viable plant. The first attempt to grow the embryos of angiosperms was made by Hannig in 1904 who obtained viable plants from in vitro isolated embryos of two crucifers Cochleria and Raphanus.

In 1924, Dietrich grew embryos of different plant species and established that mature embryos grew normally but those excised from immature seeds failed to achieve the organisation of a mature embryo.

They grew directly into seedlings, skipping the stages of normal embryogenesis and without the completion of dormancy period. Laibach demonstrated the practical application of this technique by isolating and growing the embryos of interspecific cross.

Linum perenne and L, austriacum that aborted in vivo. This led Laibach to suggest that in all crosses where viable seeds are not formed, it may be appropriate to exercise their embryos and grow them in an artificial nutrient medium.

There are two types of embryo culture:

Mature Embryo Culture

It is the culture of mature embryos derived from ripe seeds. This type of culture is done when embryos do not survive in vivo or become dormant for long periods of time or is done to eliminate the inhibition of seed germination. Seed dormancy of many species is due to chemical inhibitors or acids, mechanical resistance present in the structures covering the embryo, rather than dormancy of the embryonic tissue.

Immature Embryo Culture/Embryo Rescue

It is the culture of immature embryos to rescue the embryos of wide crosses. This is mainly used to avoid embryo abortion with the purpose of producing a viable plant. The underlying principle of embryo rescue technique is the aseptic isolation of embryo and its transfer to a suitable medium for development under optimum culture conditions.

Florets are removed at the proper time and either florets or ovaries are sterilised. Ovules can then be removed from the ovaries. The tissue within the ovule, in which the embryo is embedded, is already sterile.

For mature embryo culture either single mature seeds are disinfected or if the seeds are still unripe then the still closed fruit is disinfected. The embryos can then be aseptically removed from the ovules. Utilisation of embryo culture to overcome seed dormancy requires a different procedure.

Seeds that have hard coats are sterilised and soaked in water for few hours to few days. Sterile seeds are then split and the embryos excised. The most important aspect of embryo culture work is the selection of medium necessary to sustain continued growth of the embryo. In most cases a standard basal plant growth medium with major salts and trace elements may be utilised.

Mature embryos can be grown in a basal salt medium with a carbon energy source such as sucrose. But young embryos in addition require different vitamins, amino acids, and growth regulators and in some cases natural endosperm extracts.

Young embryos should be transferred to a medium with high sucrose concentration (8-12%); which approximate the high osmotic potential of the intracellular environment of the young embryosac, and a combination of hormones which supports the growth of heart-stage embryos (a moderate level of auxin and a low level of cytokinin). Reduced organic nitrogen as aspargine, glutamine or casein hydrolysate is always beneficial for embryo culture.

Malic acid is often added to the embryo culture medium. After one or two weeks when embryo ceases to grow, it must be transferred, to a second medium with a normal sucrose concentration, low level of auxin and a moderate level of cytokinin which allows for renewed embryo growth with direct shoot germination in many cases.

In some cases where embryo does not show shoot formation directly, it can be transferred to a medium for callus induction followed by shoot induction. After the embryos have grown into plantlets in vitro, they are generally transferred to sterile soil and grown to maturity.

Applications of embryo culture are:

Prevention of Embryo Abortion in Wide Crosses: Resignation

Successful interspecific hybrids have been seen in cotton, barley, tomato, rice, legume, flax and well known intergeneric hybrids include wheat x barley, wheat x rye, barley x rye, maize x Tripsacum, Raphanus sativus x Brassica napus.

Distant hybrids have also been obtained via embryo rescue in Carica and Citrus species. Embryo rescue technique has been successfully used for raising hybrid embryos between Actidinia deliciosa x A. eriantha and A. deliciosa x A. arguata.

Table: Resistance traits transferred to cultivated species through embryo rescue technique.

Crossing species	Resistance traits (s)
Lycopersicon esculentum x L. peruvianum	Virus, fungi and nematodes
Solanum melongena x S. khasianum (Leucinodes arbonalis)	Brinjal shoot and fruit borer
Solanum tuberosum x S.etuberosum	Potato leaf roll virus
Triticum aestivum x Thynopyrum scripeum	Salt tolerance
Hordeum satovumn x H. vulgare blotch	Powdery mildew and spot
Hordeum vulgare x H.bulbosum	Powdery mildew
Oryza satva x O.minuta	Blast (Pyricularia grisea) and
Bacterial blight	(Xanthomonas oryzae)

Production of Haploids

Embryo culture can be utilised in the production haploids or monoploids. Kasha and Kao have developed a technique to produce barley monoploids. Interspecific crosses are made with Horeum bulbosum as the pollen parent, and the resulting hybrid embryos are cultured but they exhibit H. bulbosum chromosome elimination resulting in monoploids of the female parent H. vulgare.

Overcoming Seed Dormancy

Embryo culture technique is applied to break dormancy. Seed dormancy can be caused by numerous factors including endogenous inhibitors, specific light requirements, low temperature, storage requirements and embryo immaturity. These factors can be circumvented by embryo excision and culture.

Shortening of Breeding Cycle

There are many species that exhibit seed dorm that is often localised in the seed coat and/or in the endosperm. By removing these inhibitions, seeds germinate immediately.

Seeds sometimes take up and O_2 very slowly or not at all through the seed coat, and so germinate slowly if at all, e.g. Brussels sprouts, rose, apple, oil palm and iris. H (Ilex) is important plants for Christmas decorations. Ilex embryos remain in the immature heart-shaped stage though the fruits have reached maturity.

Prevention of Embryo Abortion with Early Ripening Stone Fruits

Some species produce sterile seeds that will not germinate under appropriate conditions and eventually decay in soil e.g. early ripening varieties of peach, cherry, apple, plum. Seed sterility may be due to incomplete embryo development, which results in the death of the germinating embryo.

In crosses of early ripening stone fruits, the transport of water and nutrients to the yet immature is sometimes cut off too soon resulting in abortion of the embryo. Eg: Macapuno coconuts are priced for their characteristic soft endosperm which fills the whole nut.

These nuts always fail to germinate because the endosperm invariably rots before germinating embryo comes out of the shell. Embryo culture has been practised as a general method in horticultural crops includes avocado, peach, nectarine and plum. Two cultivars 'Goldcrest peach' and 'Mayfire nectarine' have resulted from embryo culture and commercially grown.

Embryos are Excellent Materials for in Vitro Clonal Propagation

This is especially true for conifers and members of Gramineae family.

Germination of seeds of obligatory parasites without the host is impossible in vivo, but is achievable with embryo culture.

Callus Culture

Callus is basically a more or less non-organised tumor tissue which usually arises n wounds of differentiated tissues and organs. Thus, in principle, it is a non-organised and little differentiated tissue. The cells in callus are of a parenchymatous nature. When critically examined, callus culture is not homogeneous mass of cells, because it is usually made up of two types of tissue: differentiated and non- differentiated.

Explant tissue is a differentiated tissue (roots, stem, leaves, flowers, etc.) which is used as a starting material for callus induction. These explant tissues generally show distinct planes of cell division, cell proliferation and organisation into specialised structures such as vascular systems.

If there are only differentiated cells present in an isolated explant, then dedifferentiation must take place before II division can occur. Parenchyma cells present in the explant usually undergo is differentiation. If the explant already contains meristematic tissue when isolated, then this can divide immediately without dedifferentiation taking place.

Callus formation takes place under the influence of exogenously supplied growth regulators present in the nutrient medium. The type of growth regulator requirement and its concentration in the medium depends strongly on the genotype and endogenous hormone content of an explant.

These requirements can be put into three categories:

1. Auxin alone (especially in monocotyledons).

2. Cytokinin alone.

3. Both auxin and cytokinin (carrot).

If the callus is difficult to induce, or if juvenile callus is needed, then immature embryos or seedlings or parts of these are used. It should be taken into account that type of starting material (juvenile or adult) and the original – position of the explant in the plant reflects the endogenous hormone level which has an important influence on processes such as cell division and organ and embryo formation. Many other factors are important for callus formation: genotype, composition the nutrient medium, physical growth factors (light, temperature, etc.).

Sucrose or glucose (2-4%) is usually employed as the sugar source. The effect light on callus formation is dependent on the plant species; light may be required in some cases and darkness in other cases. A temperature of 22-28 °C is normally advantageous for callus formation.

After callus induction, the callus is grown further on a new medium which referred to as sub-culturing. When sub-cultured regularly on agar medium, callus cultures will exhibit an S shaped or sigmoid pattern of growth during each passage.

There are five phases of Callus Growth:

1. Lag phase, where cells prepare to divide.

2. Exponential phase, where the rate of cell division is highest.

3. Linear phase, where cell division slows but the rate of cell expansion increases.

4. Deceleration phase, where the rates of cell division and elongation decreases.

5. Stationary phase, where the number and size of cells remain constant.

Callus growth can be monitored by fresh weight measurements, which convenient for observing the growth of cultures over time in a non-destructive manner. Dry weight measurements are more accurate than fresh weight, but method requires sacrifice of the samples. Mitotic index measurement of cell division rates require extensive sampling to reduce sample error and are not easy to perform.

Organ Culture

It is an isolated organ grown in vitro. It can be given different names depends upon the organ used as an explant. For example, meristem or shoot tip culture, root culture, nucellus culture, endosperm culture, ovule culture, a culture for production of androgenic haploids while ovule and ovary culture vitro production of gynogenic haploids. The culture of plant results in three types of in vitro culture.

Organised

The culture of whole plants (embryos, seeds) and organ has been termed as organised culture. In this, characteristic organised structure of a plants individual organ is maintained. If the organised structure is not broken down, progeny arise which are identical to the original plant material (e.g. meristem culture).

Non-organised

If cells and/or tissues are isolated from an organised part of a plant, dedifferentiate and are then cultured, a non-organised growth in the form callus tissue results. If the callus disperses into clumps of cells (aggregates) a single cells result, it is referred to as suspension culture. Non-organised culture has very low genetic stability.

Non-organised/Organised

This type of culture is intermediate between the above two types. Cells in an isolated organ or tissue first dedifferentiate and then form tissues which then re-differentiate to form organs (roots or shoots) or embryos. Thus organised structures can develop from non-organised cultures either through techniques or spontaneously. In this the progeny are often not completely identical to the original plant material.

Protoplast Culture

Isolation of Protoplasts

Protoplasts (cell without cell wall) are the biologically active and most significant material of cell. Cooking for the first time isolated protoplasts of plant tissue by using cell wall degrading enzymes viz. cellulase, hemicellulase, pectinase, and protease extracted from a saprophytic fungus Trichoderma viride. Now the protoplasts are cultured in vitro. Sterilisation of the leaf samples with sodium hypochlorite solution.

Rinsing in suitable Osmaticum ie in distilled water or MS medium adjusted to a suitable pH and buffer to maintain osmotic pressure. Plasmolysis of cells takes place by keeping the stripped leaves in 13% mannitol for 3 hours. Peeled leaves are transferred into an already sterilised enzyme solution for 12-15 hours for the facilitation of the enzyme to enter the tissue.

Isolation and purification of protoplasts takes place by filtering the enzyme solution containing protoplasts through a nylon mesh (45pm). The filtrate is centrifuged 2-3 times repeating the above steps and finally a specific concentration of protoplast suspension is prepared.

Protoplast Culture and Regeneration

From the protoplast solution of known density (about 105 protoplasts /ml) about 1ml suspension is poured on sterile and cooled down nutrient medium in Petri dishes. The plates are incubated at 25 °C in a dim white light.

The protoplast regenerates a cell wall, undergo cell division and forms callus. The callus can be subcultured. Embryogenesis begins from callus when it is placed on nutrient medium lacking mannitol and auxin. The embryo develops into the seedling s and finally into mature plants.

Anther Culture

Anther culture is the process of using anthers to culture haploid plantlets. The technique was discovered in 1964 by Guha and Maheshwari. This technique can be used in over 200 species, including tomato, rice, tobacco, barley, and geranium.

Some of the advantages which make this a valuable method for obtaining haploid plants are: the technique is fairly simple it is easy to induce cell division in the immature pollen cells in some species a large proportion of the anthers used in culture respond (induction frequency is high) haploids can be produced in large numbers very quickly.

In experiments using Datura innoxia, induction frequencies of almost 100% and a yield of more than one thousand plantlets or calluses have occurred under optimal conditions from one anther. Success can be determined within 24 hours as cells begin to divide.

Plant Tissue Culture Media

Plant tissue culture media should generally contain some or all of the following components: macronutrients, micronutrients, vitamins, amino acids or nitrogen supplements, source(s) of carbon, undefined organic supplements, growth regulators and solidifying agents. According to the International Association for Plant Physiology, the elements in concentrations greater than 05 mM.l^{-1} are defined as macroelements and those required in concentrations less than 0.5 mM.l^{-1} as microelements. It should be considered that the optimum concentration of each nutrient for achieving maximum growth rates varies among species.

Macronutrients

The essential elements in plant cell or tissue culture media include, besides C, H and O, macroelements: nitrogen (N), phosphorus (P), potassium (K), calcium (Ca), magnesium (Mg) and sulphur (S) for satisfactory growth and morphogenesis. Culture media should contain at least 25-60 mM of inorganic nitrogen for satisfactory plant cell growth. Potassium is required for cell growth of most plant species. Most media contain K in the form of nitrate chloride salts at concentrations ranging between 20 and 30 mM. The optimum concentrations of P, Mg, S and Ca range from 1-3 mM if other requirements for cell growth are provided.

Micronutrients

The essential micronutrients (minor elements) for plant cell and tissue growth include iron (Fe), manganese (Mn), zinc (Zn), boron (B), copper (Cu) and molybdenum (Mo). Iron is usually the most critical of all the micronutrients. The element is used as either citrate or tartarate salts in culture media, however, there exist some problems with these compounds for their difficulty to dissolve and precipitate after media preparation. There has been trials to solve this problem by using ethylene diaminetetraacetic acid (EDTA)-iron chelate (FeEDTA). A procedure for preparing an iron chelate solution that does not precipitate have been also developed. Cobalt (Co) and iodine (I) may be added to certain media, but their requirements for cell growth has not been precisely established. Sodium (Na) and chlorine (Cl) are also used in some media, in spite of reports that they are not essential for growth. Copper and cobalt are added to culture media at concentrations of 0.1µM, iron and molybdenum at 1µM, iodine at 5µM, zinc at 5-30 µM, manganese at 20-90 µM and boron at 25-100 µM.

Carbon and Energy Sources

In plant cell culture media, besides the sucrose, frequently used as carbon source at a concentration of 2-5%, other carbohydrates are also used. These include lactose, galactose, maltose and starch and they were reported to be less effective than either sucrose or glucose, the latter was similarly more effective than fructose considering that glucose is utilized by the cells in the beginning, followed by fructose. It was frequently demonstrated that autoclaved sucrose was better for growth than filter sterilized sucrose. Autoclaving seems to hydrolyze sucrose into more efficiently utilizable sugars such as fructose. Sucrose was reported to act as morphogenetic trigger in the formation of auxiliary buds and branching of adventitious roots.

It was found that supplements of sugar cane molasses, banana extract and coconut water to basal media can be a good alternative for reducing medium costs. These substrates in addition to sugars, they are sources of vitamins and inorganic ions required growth.

Vitamins and Myo-inositol

Some plants are able to synthesize the essential requirements of vitamins for their growth. Some vitamins are required for normal growth and development of plants, they are required by plants as catalysts in various metabolic processes. They may act as limiting factors for cell growth and differentiation when plant cells and tissues are grown Invitro The vitamins most used in the cell and tissue culture media include: thiamin (B_1), nicotinic acid and pyridoxine (B_6). Thiamin is necessarily required by all cells for growth. Thiamin is used at concentrations ranging from 0.1 to 10 mg.l^{-1}. Nicotinic acid and pyridoxine, however not essential for cell growth of many species, they are often added to culture media. Nicotinic acid is used at a concentration range 0.1-5 mg.l-1 and pyridoxine is used at 0.1-10 mg.l^{-1}. Other vitamins such as biotin, folic acid, ascorbic acid, pantothenic acid, tocopherol (vitamin E), riboflavin, p-amino-benzoic acid are used in some cell culture media however, they are not growth limiting factors. It was recommended that vitamins should be added to culture media only when the concentration of thiamin is below the desired level or when the cells are required to be grown at low population densities. Although it is not a vitamin but a carbohydrate, myo-inositol is added in small quantities to stimulate cell growth of most plant species. Myo-inositol is believed to play a role in cell division because of its breakdown to ascorbic acid and pectin and incorporation into phosphoinositides and phosphatidyl-inositol. It is generally used in plant cell and tissue culture media at concentrations of 50-5000 mg.l^{-1}.

Amino Acids

The required amino acids for optimal growth are usually synthesized by most plants, however, the addition of certain amino acids or amino acid mixtures is particularly important for establishing cultures of cells and protoplasts. Amino acids provide plant cells with a source of nitrogen that is easily assimilated by tissues and cells faster than inorganic nitrogen sources. Amino acid mixtures such as casein hydrolysate, L-glutamine, L-asparagine and adenine are frequently used as sources of organic nitrogen in culture media. Casein hydrolysate is generally used at concentrations between 0.25-1 g.l^{-1}. Amino acids used for enhancement of cell growth in culture media included; glycine at 2 mg.l-1, glutamine up to 8 mM, asparagine at 100mg.l^{-1}, L-arginine and cysteine at 10 mg.l^{-1} and L-tyrosine at 100mg.l^{-1}.

Undefined Organic Supplements

Some media were supplemented with natural substances or extracts such as protein hydrolysates, coconut milk, yeast extract, malt extract, ground banana, orange juice and tomato juice, to test their effect on growth enhancement. A wide variety of organic extracts are now commonly added to culture media. The addition of activated charcoal is sometimes added to culture media where it may have either a beneficial or deleterious effect. Growth and differentiations were stimulated in orchids , onions and carrots, tomatoes. On the other hand, an inhibition of cell growth was noticed on addition of activated charcoal to culture medium of soybean. Explanation of the mode of action of activated charcoal was based on adsorption of inhibitory compounds from the medium, adsorption of growth regulators from the culture medium or darkening of the medium. The presence of 1% activated charcoal in the medium was demonstrated to largely increase hydrolysis of sucrose during autoclaving which cause acidification of the culture medium.

Solidifying Agents

Hardness of the culture medium greatly influences the growth of cultured tissues. There are a number of gelling agents such as agar, agarose and gellan gum.

Agar-solidified medium supporting plant growth.

Agar, a polysaccharide obtained from seaweeds, is of universal use as a gelling agent for preparing semi-solid and solid plant tissue culture media. Agar has several advantages over other gelling agents; mixed with water, it easily melts in a temperature range 60-100 °C and solidifies at approximately 45 °C and it forms a gel stable at all feasible incubation temperatures. Agar gels do not react with media constituents and are not digested by plant enzymes. It is commonly used in media at concentrations ranging between 0.8-1.0%. Pure agar preparations are of great importance especially in experiments dealing with tissue metabolism. Agar contains Ca, Mg and trace elements on comparing different agar brands; Bacto, Noble and purified agar, in concern with contaminants. The author, for example reported Bacto agar to contain 0.13, 0.01, 0.19, 0.43, 2.54, 0.17% of Ca, Ba, Si, Cl, SO_4^-, N, respectively. Impurities also included 11.0, 285.0 and 5.0 mg.l1- for iron, magnesium and copper as contaminants, respectively. Amounts of some contaminants were higher in purified agar than in Bacto agar of which Mg that accounted for 695.0 mg.l^{-1} and Cu for 20.0 mg.l^{-1}.

Reduction of culture media costs is continually targeted in large-scale cultures and search for cheap alternatives provided that white flower, potato starch, rice powder were as good gelling agents as agar. It was also experienced that combination of laundry starch, potato starch and semolina in a ratio of 2:1:1 reduced costs of gelling agents by more than 70%.

Growth Regulators

Plant growth regulators are important in plant tissue culture since they play vital roles in stem elongation, tropism, and apical dominance. They are generally classified into the following groups; auxins, cytokinins, gibberellins and abscisic acid. Moreover, proportion of auxins to cytokinins determines the type and extent of organogenesis in plant cell cultures.

Auxins

The common auxins used in plant tissue culture media include: indole-3- acetic acid (IAA), in-dole-3- butricacide (IBA), 2,4-dichlorophenoxy-acetic acid (2,4-D) and naphthalene- acetic acid (NAA). IAA is the only natural auxin occurring in plant tissues There are other synthetic auxins used in culture media such as 4-chlorophenoxy acetic acid or p-chloro-phenoxy acetic acid (4-CPA, pCPA), 2,4,5-trichloro-phenoxy acetic acid (2,4,5 T), 3,6- dichloro-2-methoxy- benzoic acid (dicamba) and 4- amino-3,5,6-trichloro-picolinic acid (picloram).

Auxins differ in their physiological activity and in the extent to which they translocate through tissue and are metabolized. Based on stem curvature assays, eight to twelve times higher activity was reported on using 2,4-D than IAA, four times higher activity of 2,4,5 T in comparison with IAA and NAA has as doubled activity as IAA. In tissue cultures, auxins are usually used to stimulate callus production and cell growth, to initiate shoots and rooting, to induce somatic embryogenesis, to stimulate growth from shoot apices and shoot stem culture. The auxin NAA and 2,4-D are considered to be stable and can be stored at 40C for several months. The solutions of NAA and 2,4-D can also be stored for several months in a refrigerator or at -200C if storage has to last for longer periods. It is best to prepare fresh IAA solutions each time during medium preparation, however IAA solutions can be stored in an amber bottle at 40C for no longer than a week. Generally IAA and 2,4-D are dissolved in a small volume of 95% ethyl alcohol. NAA, 2,4-D and IAA can be dissolved in a small amount of 1N NaOH. Chemical structures of some of the frequently used auxins are given in figure.

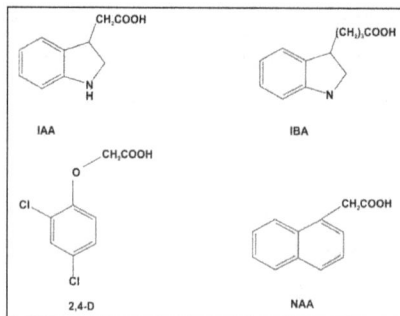

Chemical structure of commonly used auxins. IAA indole acetic acid,
IBA Indole butyric acid, 2,4-D dichlorophenoxyacetic acid and
NAA naphthalene acetic acid.

There are also some auxinlike compounds that vary in their activity and are rarely used in culture media.

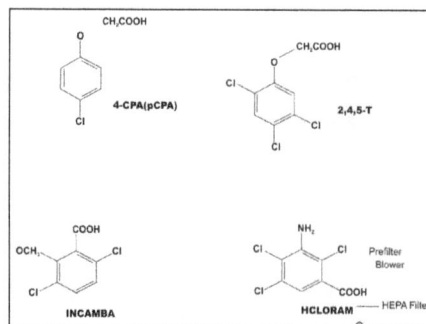

Some auxinlike compounds. pCPA, p-chloro-phenoxy acetic acid
and 2,4,5T, 2,4,5-trichloro-phenoxy acetic acid.

Cytokinins

Cytokinins commonly used in culture media include BAP (6-benzyloaminopurine), 2iP (6-dime-thylaminopurine), kinetin (N-2-furanylmethyl-1H-purine-6-amine), Zeatin (6-4-hydroxy-3-meth-yl-trans-2-butenylaminopurine) and TDZ (thiazuron-N-phenyl-N-1,2,3 thiadiazol-5ylurea). Zeatin and 2iP are naturally occurring cytokinins and zeatin is more effective. In culture media, cytokinins proved to stimulate cell division, induce shoot formation and axillary shoot proliferation and to retard root formation. The cytokinins are relatively stable compounds in culture media and can be stored desiccated at -20 °C. Cytokinins are frequently reported to be difficult to dissolve and some-times addition of few drops of 1NHCl or 1N NaOH facilitate their dissolution. Cytokinins can be dissolved in small amounts of dimethylsulfoxide (DMSO) without injury to the plant tissue. DMSO has an additional advantage because it acts as a sterilizing agent; thus stock solutions containing DMSO can be added directly to the sterile culture medium. Chemical structure of the frequently used in plant tissue culture media is given in figure.

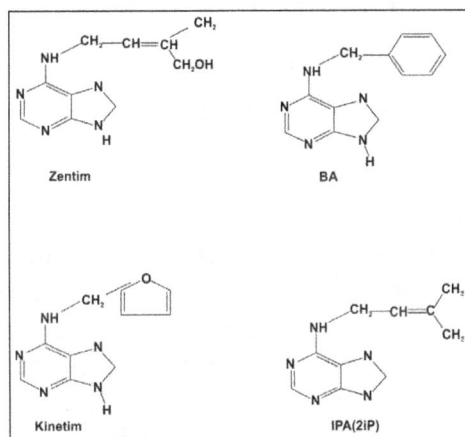

Chemical structure of some cytokinins, BA, benzyladenine,
IPA dimethylallylamino purine.

Gibberellins

Gibberellins comprise more than twenty compounds, of which GA_3 is the most frequently used gibberellin. These compounds enhance growth of callus and help elongation of dwarf plantlets.

Other growth regulators are sometimes added to plant tissue culture media as abscisic acid, a compound that is usually supplemented to inhibit or stimulate callus growth, depending upon the species. It enhances shoot proliferation and inhibits later stages of embryo development. Although growth regulators are the most expensive medium ingredients, they have little effect on the medi-um cost because they are required in very small concentrations.

Media Preparation

Preparation of culture media is preferred to be performed in an equipped for this purpose com-partment. This compartment should beconstructed so as to maintain ease in cleaning and reduc-ing possibility of contamination. Supplies of both tap and distilled water and gas should be provid-ed. Appropriate systems for water sterilization or deionization are also important. Certain devices are required for better performance such as a refrigerator, freezer, hot plate, stirrer, pH meter, electric balances with different weighing ranges, heater, Bunsen burner in addition to glassware and chemicals. It is well known now that mistakes which occur in tissue culture process most frequently originate from inaccurate media preparation that is why clean glassware, high quality water, pure chemicals and careful measurement of media components should be facilitated.

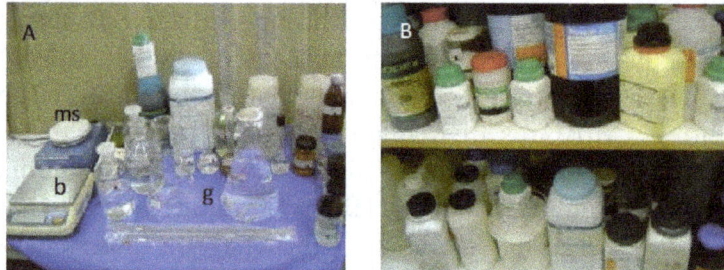

Some of components of the preparation room. A, some equipments used; ms magnetic stirrer hot plate, b electric balance, g glassware. B shelves for keeping chemicals.

A convenient method for preparation of culture media is to make concentrated stock solutions which can be immediately diluted to preferred concentration before use. Solutions of macronutrients are better to be prepared as stock solutions of 10 times the strength of the final operative medium. Stock solutions can be stored in a refrigerator at 2- 4 °C. Micronutrients stock solutions are made up at 100 times of the final concentration of the working medium. The micronutrients stock solution can also be stored in a refrigerator or a freezer until needed. Iron stock solution should be 100 times concentrated than the final working medium and stored in a refrigerator. Vitamins are prepared as either 100 or 1000 times concentrated stock solutions and stored in a freezer (-20 °C) until used if it is desired to keep them for long otherwise they can be stored in a refrigerator for 2-3 months and should be discarded thereafter. Stock solutions of growth regulators are usually prepared at 100-1000 times the final desired concentration.

Concentrations of inorganic and organic components of media are generally expressed in mass values (mg.l^{-1}, mg/l and p.p.m.) in tissue culture literature. The International Association for plant Physiology has recommended the use of mole values. Mole is an abbreviation for gram molecular weight which is the formula weight of a substance in grams. The formula weight of a substance is equal to the sum of weights of the atoms in the chemical formula of the substance. One liter of solution containing 1mole of a substance is 1 molar (1M) or 1 mol.l-1 solution of the substance (1 mol.l^{-1}= 10^3 mmol.l^{-1}= 10^6 µmol/l). It is routinely now to accepted to express concentrations of macronutrients and organic nutrients in the culture medium as mmol/l values, and µmol/l values for micronutrients, hormones, vitamins and other organic constituents. This was explained on the basis that mole values for all compounds have constant number of molecules per mole.

Media Selection

For the establishment of a new protocol for a specific purpose in tissue culture, a suitable medium is better formulated by testing the individual addition of a series of concentrations of a given compound to a universal basal medium such as MS, LS or B$_5$. The most effective variables in plant tissue culture media are growth regulators, especially auxins and cytokinins. Full strength of salts in media proved good for several species, but in some species the reduction of salts level to / or / the full concentration gave better results in in vitro growth.

Sucrose is often assumed to be the best source of carbon for in vitro culture, the levels used are from 2 to 6% and the level has to be defined for each species.

Media Sterilization

Prevention of contamination of tissue culture media is important for the whole process of plant propagation and helps to decrease the spread of plant parasites. Contamination of media could be controlled by adding antimicrobial agents, acidification or by filtration through microporous filters. To reduce possibilities of contamination, it is recommended that sterilization rooms should have the least number of openings. Media preparation and sterilization are preferred to be performed in separate compartments. Sterilization area should also have walls and floor that withstand moisture, heat and steam.

Sterilization of media is routinely achieved by autoclaving at the temperature ranging from 115 °C – 135 °C. Advantages of autoclaving are: the method is quick and simple, whereas disadvantages are the media pH changes and some components may decompose and so to loose their effectiveness. As example autoclaving mixtures of fructose, glucose and sucrose resulted in a drop in the agar gelling capacity and affecting pH of the culture medium through the formation of furfural derivatives due to sucrose hydrolysis.

Filtration through microporus filters (0.22-0.45) is also used for thermolabile organic constituents such as vitamins, growth regulators and amino acids. Additives of antimicrobial agents are less commonly applied in plant tissue culture media. Limitation for their use was reported and attributed to harm imposed on plants as well.

Plant Tissue Culture Tools and Techniques

The applications and principle of each of the tools and techniques used for plant tissue culture are briefly described hereunder with schematic diagrams.

pH and pH Meter

The hydrogen ion concentration of most solutions is extremely low. pH of a solution is strictly defined as the negative logarithm of the hydrogen ion activity. But in practice usually hydrogen ion concentration is taken.

$$pH = -\log_{10}(H^+) = 7$$

The pH of pure water is 7 at 25 °C. Generally glass distilled water is used for the preparation of culture medium. However, sometimes buffered solutions may be used for the same to keep the pH of the medium constant.

Measurement of pH

An approximate idea of the pH of a solution can be obtained using indicators. These are organic compounds of natural or synthetic origin whose Colour is dependent upon the pH of the solution.

Indicators are usually weak acids, which dissociate in solution. A standard pH meter has two electrodes, one glass electrode for measuring pH and the other calomel reference electrode. Reference electrode is filled with saturated KCI solution.

$$\text{Indicator} = \text{Indicator}^- + H^+$$

The pH probe measures pH as the activity of hydrogen ions surrounding a thin walled glass bulb at its tip. The probe produces a small voltage (about 0.06 volt per pH unit) that is measured and displayed as pH units by the meter. The meter circuit is fundamentally no more than a voltmeter that displays measurements in pH units instead of volts.

The input impedance of the meter must be very high because of the high resistance approximately 20 to 1000 MΩ of the glass electrode probes typically used with pH meters. The circuit of a simple pH meter usually consists of operational amplifiers in an inverting configuration, with a total voltage gain of about $- 17$. The inverting amplifier converts the small voltage produced by the probe (+ 0.059 volt/pH) into pH units, which are then offset by seven volts to give a reading on the pH scale.

For example:

1. At neutral pH (pH 7) the voltage at the probe's output is 0 volts.

2. At alkaline pH, the voltage at the probe's output ranges from + 0 to + 0.41.

3. At acid pH, the voltage at the probe's output ranges from $- 0.41$ volts to $- 0$.

4. Now-a-days in more sophisticated pH meters, the both types of the electrodes are combined in one.

Applications

1. To adjust pH of different solutions, preparation of buffers and culture media.

2. Determination of pH of cells (cell sap) and in analytical techniques.

3. To monitor pH of the medium in a bioreactor.

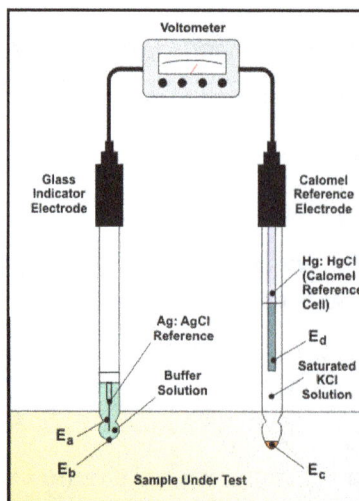

The construction of pH meter electrodes and basic principle.

Autoclave

Autoclave is used to sterilize medium, glassware and tools for the purpose of plant tissue culture. The same equipment is used in hospitals to sterilize gauge, cotton, tools and linen, etc. Sterilization of material is carried out by increasing moist heat (121 °C) due to increased pressure inside the vessel (15-22 psi, pounds per square inch or 1.02 to 1.5 kg/cm^2) for 15 minutes for routine sterilization. Moist heat kills the microorganism and makes the material free from microbes.

Construction

Autoclaves of different sizes from 5 litres to several hundred litres capacity are available in horizontal or vertical designs. As shown in figure, an autoclave have a body, an internal (or external for small sized autoclaves comparable to household pressure cooker) heating system, a container to hold material, its cover fixed with pressure gauge, safety valve, pressure release valve etc. Lid is tightened with the help of screws and a gasket seals the body and lid. A jacket, paddle lifter, timer, and indicator etc., are also provided with large sized autoclaves. Autoclaves may be constructed of aluminum, mild steel, stainless steel or gun metal. Industrial autoclave can accommodate large trolley containing huge number of glassware's or large bioreactors.

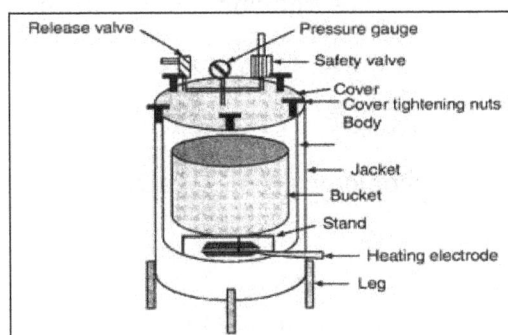

A schematic diagram of laboratory autoclave.

Operation

Place the materials (wrapped in aluminum foil or paper or in metal box) and glass-wares containing medium (plugged with non-absorbent cotton and covered with aluminum foil) in the bucket. Check water level for appropriate level, tighten all the screws, and switch-on the current. Allow the steam to pass freely from release valve for 5 minutes and then close the valve.

After attaining a pressure of 15 psi, count 15 minutes for sterilization and then switch-off the current. Pressure is maintained by safety valve. Modern autoclaves are fitted with temperature and time control units and can automatically control the period for sterilization and then switch-off themselves.

Empty vessels, beakers, graduated cylinders, etc., should be closed with a cap or aluminum foil. Tools should also be wrapped in foil or paper or put in a covered sterilization tray. It is critical that the steam penetrate the items in order for sterilization to be successful.

For large sized vessels and large volume flasks containing high amount of liquid, duration for sterilization should be increased accordingly. Table shows the relationship between volume of the solution and duration for its sterilization at 15 psi. The vessels should not be filled more than 1/3 of its capacity for proper sterilization.

Volume of Solution and Duration for its Sterilization

ml/Container	Duration at 121 °C (min)
20-50	15
75	20
250-500	25
1000	30
1500	35
2000	40

Precautions:

1. Check water level each time, the heating elements should remain immersed in the water.

2. Check spring of safety valve frequently and clean opening whenever necessary.

3. Opposite screws of the lid should be tightened simultaneously.

4. Do not over tighten the screws to avoid damage to the gasket.

5. Use permanent marker to mark your flasks and its medium.

6. All the electrical equipment should be properly earthen to avoid electric shock.

Plant Growth Chamber

Plant growth chambers can be constructed in a suitable sized room or can be purchased as commercially available equipment. Thermal insulation of walls increases the efficiency of the cooling system.

Essentially plant growth chamber has three environmental control systems:

1. Light-intensity and duration cycle control.

2. Temperature control and regulation.

3. Humidity control and regulation.

All the modern instruments are electronically controlled precision instruments with sophisticated sensors and timers to regulate the desired set values.

Light

Light is fixed in the roof of equipment or in shelves. Light is provided by commercially available light sources like cool white fluorescent tubes and incandescent lamps in a ratio of 3:1 and usually a light intensity of 2000-2500 lux (about 200-250 candles or 30 μ mol m-2 s1) is used. The duration of light and dark cycle is adjusted as per requirement, usually 16 hours light cycle is given. Nowadays warm fluorescent tubes are also available which provides wide spectrum as compared to cool white fluorescent tubes and mixing of incandescent light is not required with former tubes. Light intensity can also be regulated by photoperiod simulators. Thus, light quality, intensity and period is controlled and regulated by the instrument as per set valves.

Temperature

In modern equipment, temperature is precisely regulated by good quality (platinum) temperature sensor. In all cases, air conditioning units provide the cooling. It is always advisable to keep one spare compressor unit, for emergency, to avoid delay in repairs and damage to cultures. Usually temperature of 22-28 °C is used for growing plant tissue culture. Temperatures should be measured in a constructed growth chamber at different levels and places, viz., light racks, central and corners to have a correct temperature setting.

Humidity

Humidity inside the growth chamber is provided by humidifier (a mist generating system) and controlled by humidistat. Usually 60% RH (relative humidity) is used to maintain healthy growth. Low RH may cause early drying of medium while high humidity may cause fungal growth in the environment and on a various articles. Thus, in a growth chamber, light, temperature and humidity are precisely controlled and cultures are grown in a controlled environment. All the controls are set on control panel.

Laminar Air Flow Bench

Laminar air flow (LAF) bench is the main working table for aseptic manipulations related to plant tissue culture. This is equipment fitted with High Efficiency Particulate Air (HEPA) Filters, which allow air to pass but retain all the particles and micro-organisms. These HEPA Filters have a very small pore size (0.3 μm) with 99.97-99.99% efficiency.

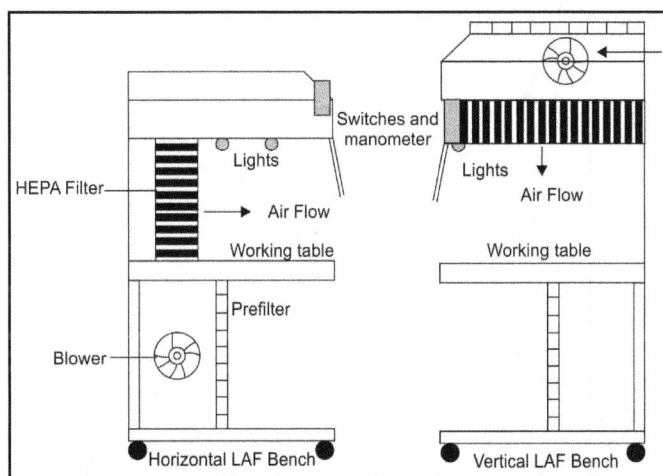

Line diagram of horizontal laminar air flow bench.

Air blown by a blower is passed through these filters, thus, always generally a positive pressure of air is maintained from inside to outside. This positive pressure does not allow any particle to enter in the working area of LAF bench. Air pressure inside the instrument is measured by a manometer and pressure of more than 12 bars shows choking of filters and at this point filters should be replaced. Equipment is fitted with UV light and visible light source.

UV is switched-on for 30 minutes before starting the work to make area free from microbes. LAF Bench of steel and wooden cabinets is available with different working table size and for vertical (downward air flow) or horizontal (horizontal flow) model. There are several manufacturers in India for this instrument (body) but HEPA filters are imported (Fig.). HEPA filters are also used to create 'clean area' for culture rooms and inoculation room etc. If LAF bench is placed in such a clear area, efficiency and life of the equipment are increased.

Microscopy

Electron Microscopy

Electron microscopy permits a detailed study of sub-cellular organelles as its resolving power is much greater than that of the light microscope. Max Knoll and Ernst Ruska in 1931, at Technical University in Berlin, constructed electron microscope (EM). In the EM, streams of electrons are deflected by an electrostatic or electromagnetic field in the same way that a beam of light is refracted by a lens.

Electron beam is generated by heating a filament in vacuum, which are accelerated by a potential and shows properties similar to light (λ = 0.005nm of electrons and 550 nm for light). Though appears to be similar there are great differences in light and electron microscopes, the principal being the image formation. In electron microscope, image is produced by electron scattering.

Electron dispersion is a function of the thickness and molecular packing of the object and depends especially on the atomic number of the atoms in the object. The higher the atomic number, the greater is the dispersion. There are three general types of electromagnetic lenses.

The one is placed between the source of illumination and the specimen. This focuses the beam of electrons on specimen and functions in a similar manner as the condenser in light microscope. The other two lens systems are on the opposite side of specimen which magnifies the image in similar way as objective and ocular in light microscope.

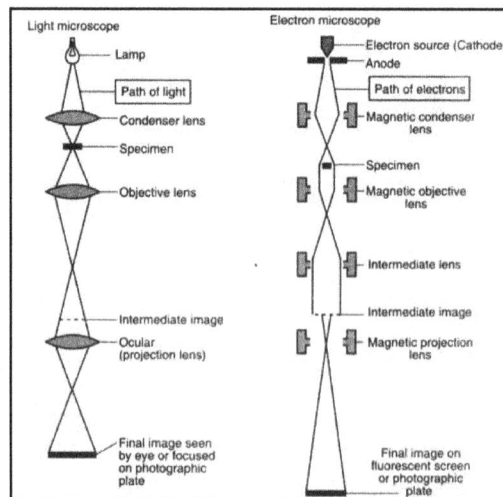

Diagrammatic comparison of image formation in electron microscope and light microscope.

The final image is produced on a screen coated with a phosphorus compound which fluoresces upon irradiation by electrons. These images are recorded on photographic films. Air molecules in the microscope interfere with the movement of electrons. To prevent this, a high vacuum (10^{-4} – 10^{-6} mm Hg) is created inside the microscope.

In light microscope, magnification is largely determined by the objective and the maximum magnification of 100-120 can be obtained. This is multiplied by ocular lens by a factor of 5-15, reaching a total magnification of 500-1500x.

The resolving power of Transmission Electron Microscope (TEM) is very high. The image generated by objective can be multiplied several hundred times by projector coil, e.g., 100 objective 200x projector coil = 20,000x. In TEM, this can be reached up to 10,000,000 xs. Electron microscope has a greater depth of field as compared to light microscope. This property of EM is used to develop three dimensional images of cell organelles as ribosome's and protein structure.

The preparation of specimen is similar to dehydration used for histology for light microscopy. The sections prepared for EM should be thin (less than 0.5 or 500 nm), otherwise, material will appear completely dark. Such thin sections (50-300 nm) are prepared with the help of ultra-microtome, fitted with glass or diamond knife. To cut sections, material is fixed in osmium tetra oxide, processed through a dehydration series and embedded in plastic or resin (araldite).

Sections are placed on small grids specifically used for EM. It is difficult to handle ultra thin sections so they are placed on copper grids (400 apertures/inch square).The grids are placed in labeled boxed and can be stored till observed in EM and can be used again and again without damaging the material. Dehydrated materials are placed in vacuum chamber and vapour of heavy metals like platinum, chromium or palladium is directed from a filament of incandescent tungsten.

The vaporized material is deposited on one side of the surface of elevated particles. This creates a shadow on the other side (Shadow casting). Height of the material is determined by length of shadow. In photographs, such specimens appear in three dimensions, which is not possible with other techniques.

Freeze fracturing (fracture of frozen material) is used to see fine details of replicated material. These are different techniques to prepare materials for sectioning and for staining to increase contrast. Special techniques are used for bacteria, viruses, proteins and nucleic acids. TEM is used to study the fine details of a cell, compare the clones, development of plastids and cell wall formation.

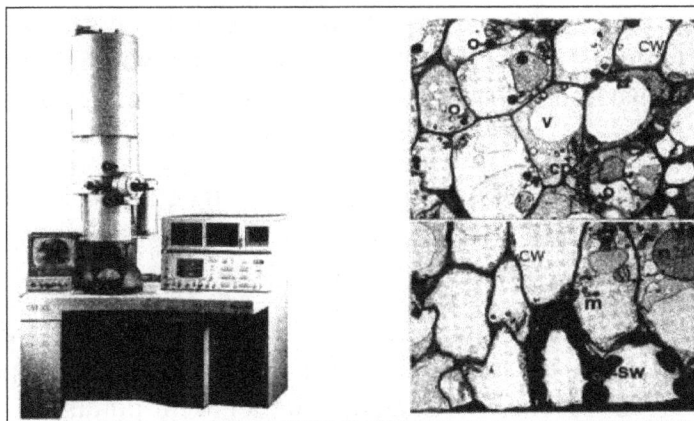

TEM photos showing primary cell wall (cw), secondary cell wall (sw), mitochondria (m), chloroplasts (cp), nucleus (n) and oil droplets (o).

SEM

Scanning electron microscope (SEM) provides surface views of whole structure of specimen. Normally, specimens are coated with a thin film of metal under vacuum. As compared to shadow casting, complete coating is essential because the scanning beam produces a charge on the uncoated biological materials which cause distortion of image.

The metal coated specimen is scanned with a beam of electrons (50-100 Å) which strikes on the specimen and emits the secondary electrons. These electrons pass through a cathode tube and image is produced on the screen of Cathode Ray Tube. In plant system, SEM is used to see the nature of appendages, pollen morphology and morphogenesis in plant tissue culture. It is also used to study the structures of metals, crystals and in forensic sciences.

SEM photograph of semiorganized and fractured callus.

Light Microscopy

Bright field microscopy is absolutely indispensable tool for cell biologists. This is required for routine observations of cells, cellular differentiation and pigmentation. Fluorescence microscopy has become a powerful tool for cell biologists, particularly for the selection of fluorescing secondary metabolite rich cells. Only these two techniques are discussed here in brief.

According to wave theory, light is propagated from one place to another as wave travelling in a hypothetical medium. Light waves are described in terms of their amplitude, frequency, and wavelength. Amplitude is the maximum displacement of light path from the position of equilibrium.

Frequency of light is the number of complete cycles occurring in a second. The property of light waves inversely associated with the frequency is the wavelength and is defined as the distance between corresponding points on a wave or distance between two successive peaks or crests. The velocity of light is about 186300 miles or 3 x 1015 kms per second.

A good microscope has not only good magnifying power but also good resolving power to provide finer details of the object. Thus, the basic difficulty in designing a microscope is riot the magnification, but the ability of lens system to distinguish two adjacent points as distinct and separate. This ability is known as resolving power of the microscope. The resolving power of a microscope depends upon the wavelength of light and numerical aperture (N.A.).

The minimum resolvable distance between two luminous points (v) is given by the following formula:

$$v = \frac{0.16\lambda}{\text{N.A.}} \text{ where, } \lambda = \text{wavelength}$$

Thus, shorter the wavelength of light used and lower the N.A., greater is the resolving power. The limit of resolution of a microscope is approximately equal to 0.5 /N.A., which for a light microscope is approximately 200 nanometers (nm) or about the size of many bacterial cells.

The numerical aperture of a lens is dependent on the refractive index (r) = the ratio of the speed of light in a given medium to the speed of light in a vacuum) of the medium filling the space between the specimen and the front of the objective lens and on the angle of the most oblique rays of light that can enter the objective lens (θ). It is given by formula [N.A. = ηx sinθ], that is why immersion oil is placed between the object and oil immersion lens (100 X objective).

Colorimeter

The most commonly used method for determining the concentration of biochemical compounds is colorimetry. It uses the property of light such that when white light passes through a coloured solution, some wavelengths are absorbed more than others. Hyaline solution can be made coloured by specific reactions with suitable reagents. These reactions are generally very sensitive to determine quantities of material in the region of millimole per litre concentration.

The big advantage is that complete isolation of the compound is not necessary and the constituents of a complex mixture such as blood can be determined after little treatment. The depth of colour is directly proportional to the concentration of the compound being measured, while the amount of light absorbed is proportional to the intensity of the colour and therefore, to the concentration.

Source of Radiation

A lamp is usually used as the source of radiation. The wavelength of light depends upon the quality of source lamp. Colorimetry is an analytical device to determine the amount of an unknown substance in a solution. The device works on optical properties of the substance. Basically we measure the amount of light of a particular wavelength absorbed by that solution, and then relate solute concentration to the absorbance.

A simplified schematic presentation of the instrument is as follows:

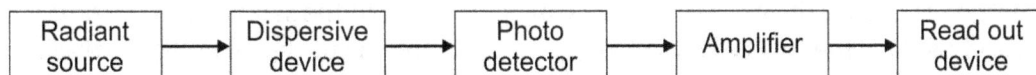

Radiant source	→	Dispersive device	→	Photo detector	→	Amplifier	→	Read out device

The relationship between concentration of solution and light it absorbs is given by the following laws.

Lambert's Law

When a ray of monochromatic light passes through an absorbing medium its intensity decreases exponentially as the length of the absorbing medium increases.

Beer's Law

When a ray of monochromatic light passes through an absorbing medium its intensity decreases exponentially as the concentration of the absorbing medium increases.

Transmittance

The ratio of intensities is known as the transmittance and this is usually expressed as percentage. Absorbance or extinction is known as the optical density of the substance. Absorbance increases with increase in concentration of the solution.

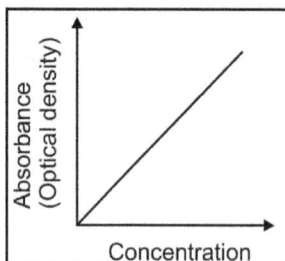

Relationship between concentration and absorbance.

Detection of Radiation

Detection wavelength is selected on the basis of wavelength of light coming out of a solution. Absorption of a particular wavelength of light is characteristic of a compound. Outgoing radiations are detected by photochemical detectors or vacuum phototubes. If the light intensity is very low, a photomultiplier tube is used. A phototube converts the transmitted light energy into an electric current. This electric signal is calibrated and can be produced on a chart recorder.

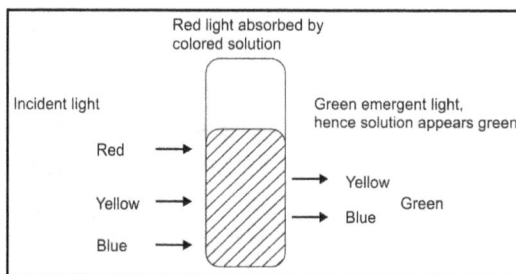

Solution appears coloured depending upon
the wavelength of the outgoing light.

A better method is to split the light beam, pass one part through the sample and the other through the blank, and balance the two circuits to give zero (a double beam system). The extinction is determined from the potentiometer reading, which balances the circuit.

Applications

Measurement of Concentrations

Spectrophotometers are widely used for quantitative measurements of substance concentration in the solutions by measuring absorbance at the optimal wavelength. This is used to determine concentration in fractions obtained in column chromatography, TLC and in other solutions.

Absorption Spectra

Identification and structure evaluation of various isolated pure compounds can be done by UV-Visible spectrophotometry. All the compounds have very specific absorption spectra depending on the molecular structure. Comparison of spectra helps in identification of compounds. The wavelength at which maximum absorption takes place is represented as λ_{max} of that compound in a given solvent.

Enzyme Kinetics

In enzymatic reactions, changes in absorbance are recorded to determine the reaction velocity and concentration of the product formed/substrate utilized. Spectrophotometers coupled with microprocessor or computer can store data, figures and are helpful in comparing the effect of different variables.

Centrifugation

A centrifuge is an instrument which produces centrifugal force by rotating the samples around a central axis with the help of an electric motor. Centrifuges can be categorized as the clinical type (5-10,000 rpm), refrigerated high-speed centrifuges (10,000-20,000 rpm) and ultra-centrifuges (20,000 to 80,000 rpm).

With increase in rpm, the friction of rotor with air produces so much of heat that they have to be run under refrigeration (so called refrigerated centrifuge) and both refrigeration and vacuum are used in ultracentrifuge, which runs at very high rpm. For these high speeds, even the rotor has to be made of special metal to withstand the great force.

There are two types of rotors, angle head and swing. In the former, the samples are kept at an angle of about 30° to the vertical axis whereas; in the latter the samples while spinning are horizontal. Simple calculations show that for the same radius, the swinging bucket method produces more gravitational force. Ultracentrifuges are of two types – analytical and preparative model.

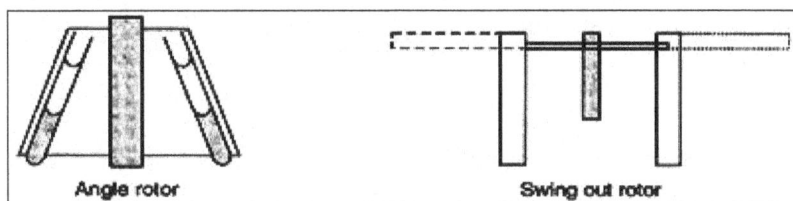

Different types of rotors used in centrifugation.

Analytical Model

This consists of rotors and tubes, called cells. The instrument is designed to allow the operator to follow the progress of the substances in the cells, while the process of centrifugation is in progress. By estimating sedimentation velocity during the process, the molecular weight, purity etc. can be determined.

Preparative Model

This is used for purification of the components of macromolecules or other substances and all determinations are made at the end of centrifugation. The instrument has no monitoring device, while large centrifugal forces are set for a fixed time period. Centrifugation is the most widely used technique for separation of various metabolites and also used to separate non-miscible liquids during extraction of secondary metabolites, e.g., water (aqueous), chloroform (organic) mixture.

A wide variety of centrifuges are available, ranging in capacity and speed. During the process of centrifugation, solid particles experience a centrifugal force, which pulls them outwards, i.e., away from the center. The velocity with which a given solid particle moves through a liquid medium is related to angular velocity. The principle of centrifugation is that, an object moving in a circular motion at an angular velocity is subjected to an outward force (F) through a radius of rotation (r) in cms, presented as $\omega^2 [F = \omega^2 r]$.

F frequently expressed in terms of gravitational force of the Earth, commonly referred to as RCF (Relative centrifugal force) or g by the following formula $[RCF = \omega 2r / 980]$. The ω^2 operating speed of the centrifuge is expressed as revolutions per minute 'rpm', which can be converted to radians by the following formula:

$$\omega = \frac{rpm}{30} \frac{\pi (rpm)}{30}$$

or,
$$RCF = \frac{\frac{(\pi\,rpm)^2 (r)}{30^2}}{980} = \frac{980(\pi rpm)^2 r}{30^2}$$

Therefore, $RCF = 1.2 \times 10^{-5} (rpm)^2 r$

Sometimes the velocity of the moving particles is expressed in the form of sedimentation coefficient (s) by the formula $[V - s\,(\omega^2 r)]$. The sedimentation coefficient is a characteristic constant for a molecule or a particle and is a function of the size, shape and density. It is equivalent to the average velocity per unit of acceleration. The unit Svedberg (s) is often used with reference to centrifugation and is equivalent to a sedimentation coefficient of 10^{-13} s.

Applications

Centrifugation is widely used in analytical techniques, preparation of extracts, separation of non-miscible liquid mixtures, purification of enzymes, inhibitors and removing particles (silica, etc.) from the solvents. Centrifugation with heating and vacuum (suction) is used for concentrating extraction (instrument called sample concentrator) and rapidly removing solvents. All the enzymatic work requires refrigerated centrifuges to keep the samples cool during centrifugation and to protect enzymes from inactivation by heat.

Chromatography

Chromatography (meaning 'colored writing') is a technique to separate molecules on the basis of differences in size, shape, mass, charge and adsorption properties. The term chromatography was used by the Russian botanist Tswett to describe the separation of plant pigments on a column of alumina. There are different types of chromatography but they all involve interactions between these components: the mixture to be separated, a solid phase, and a solvent.

The magnitude of these interactions depends upon the particular method used. Usually column chromatography is used to separate large quantities of compounds, whereas, paper chromatography (PC) or thin-layer chromatography (TLC) in one or two dimensions, is used for analytical work.

The mobile phase can be a gas or a liquid, whereas the stationary phase can only be a liquid or solid. In liquid column chromatography (LCC), separation involves a simple partitioning between two immiscible liquid phases, one stationary and the other mobile, the process is called liquid-liquid (or partition) chromatography (LLC). When physical surface forces are mainly involved in the retentive ability of the stationary phase, the process is denoted liquid-solid (or adsorption) chromatography (LSC).

In ion exchange chromatography (IEC), ionic components of the sample are separated by selective exchange with counter ions of the stationary phase. Use of exclusion packing's as the stationary phase brings about a classification of molecules based largely on molecular geometry and size.

Paper Chromatography

Principle

Cellulose in the form of paper sheets makes an ideal support medium where water is absorbed between the cellulose fibres and forms a stationary hydrophilic phase. The suitably concentrated mixture is spotted onto the paper, dried with a hair-drier and the chromatogram is developed by allowing the solvent to flow along the sheet. The solvent front is marked and after drying the paper, the positions of the compounds present in the mixture are visualized by a suitable staining reaction.

The ratio of the distance moved by a compound to that moved by the solvent is known as the R; value and is more or less constant for a particular compound, solvent system and paper under carefully controlled conditions of solute concentration, temperature and pH.

Sample

Generally, alcoholic extracts of plant material with or without partial purification are used for chromatography. Biological materials should be desalted before chromatography by electrolysis or electro-dialysis. Excess salt results in a poor chromatogram with spreading of spots and changes in their R. values.

It can also affect the chemical reactions used to detect the compounds being separated. The sample (10-20 µl) is then applied to the paper with a micropipette or capillary.

Paper

Whatman No. 2 is the paper most frequently used for analytical purposes. Whatman No. 3 MM is a thick paper and is best employed for separating large quantities of material; the resolution is, however, inferior to Whatman No. 1. For rapid separation, Whatman Nos. 4 and 5 are convenient, although the spots are less well defined. In all cases, the flow rate is faster in the 'machine direction', which is normally noted on the box containing the paper.

The paper may be impregnated with a buffer solution before use or chemically modified by acetylation. Ion exchange papers are also available commercially. For the separation of lipids and similar hydrophobic molecules, silica-impregnated papers are available commercially.

Solvent

This choice, like that of the paper, is largely empirical and will depend on the mixture investigated. If the compounds move close to the solvent front in solvent 'A' then they are too soluble, while if they are crowded around the origin in solvent 'B' then they are not sufficiently soluble.

A suitable solvent for separation would, therefore, be an appropriate mixture of 'A' and 'B', so that the R_f values of the components of the mixture are spread across the length of the paper.

R_f value is defined as [R_f = the distance moved by solute / the distance moved by solvent front]. This value is a constant for a particular compound under standard conditions and closely reflects the distribution coefficient for that compound. It is recommended that the developing chamber should be saturated with solvent, by using filter paper lining inside the chamber.

Two Dimensional Chromatography

The mixture is separated in the first solvent, which should be volatile, then after drying, the paper is turned through 90° and separation is carried out in the second solvent. After location, a map is obtained and compounds can be identified by comparing their position with a map of known compounds developed under the same conditions.

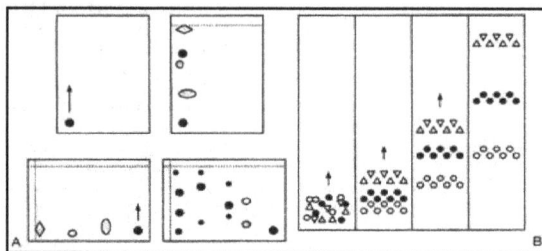

Separation of a mixture by (A) two-dimensional chromatography
and (B) principle behind the chromatographic separation.

Detection of Spots

Most compounds are colourless and are visualized by specific reagents. The location reagent is applied by spraying the paper under a fumigation hood or rapidly dipping it in a solution of the reagent in a volatile solvent. Viewing under ultraviolet light is also useful since some compounds, which absorb strongly show up as dark spots against the fluorescent background of the paper. Other compounds show a characteristic fluorescence under ultraviolet light.

Thin Layer Chromatography

Principle

Separation of compounds on a thin layer of adsorbing material is similar in many ways to paper chromatography, but has the added advantage that a variety of supporting media can be used so that separation can be by adsorption ion exchange, partition chromatography, or gel filtration depending on the nature of the medium employed.

The method is very rapid compared to paper chromatography and many separations can be completed within an hour. Compounds can be detected at a lower concentration than on paper as the spots are very compact. Furthermore, separated compounds can be detected by corrosive sprays and elevated temperatures with some thin layer materials, which of course is not possible with paper.

Production of Thin Layer

The R_f value is affected by the thickness of the layer below 200 μm and a thickness of 250 μm is suitable for most separations. There are several good spreaders available in the market, which can produce an even layer of required thickness by adjusting the thickness control screw.

Calcium sulphate is sometimes incorporated into the adsorbent to bind the layer to the plate and, because of this; it is advisable to work rapidly once the adsorbent is mixed with water.

There are now a number of prepared thin layer plates using different adsorbents on various supporting materials such as glass, plastic, and aluminum that are available commercially and these may be more convenient to use than trying to prepare plates in the laboratory.

Development

It is essential to make sure that the atmosphere of the separation chamber is fully saturated with the solvent mixture, otherwise R_f values will vary widely from tank to tank. However, horizontal chambers of high performance TLC (HPTLC) are very useful and convenient as they require less time, solvent and no stabilization time. Development of the plate is usually by the ascending technique and is very rapid.

Thermometer

The thermometer is a device that measures temperature or temperature gradient using a variety of different principles; A thermometer has two important elements: the temperature sensor (e.g., the bulb on a mercury thermometer) in which some physical change occurs with temperature, plus some means of converting this physical change into a value (e.g., the scale on a mercury thermometer).

Industrial thermometers commonly use electronic means to provide a digital display or input to a computer. The Alcohol thermometer or Spirit thermometer is an alternative to the Mercury-in-glass thermometer, and functions in a similar way. An organic liquid is contained in a glass bulb which is connected to a capillary of the same glass and the end is sealed with an expansion bulb. The space above the liquid is a mixture of nitrogen and the vapour of the liquid.

For the working temperature range, the meniscus or interface between the liquid is within the capillary. With increasing temperature, the volume of liquid expands and the meniscus moves up the capillary. The position of the meniscus shows the temperature against an inscribed scale.

The liquid used can be pure ethanol or toluene or kerosene or Isoamyl acetate, depending on manufacturer and working temperature range. Since these are transparent, the liquid is made more visible by the addition of a red or blue dye. One half of the glass containing the capillary is usually enameled white or yellow to give a background for reading the scale.

Temperature is measured by maximum-minimum thermometer or a continuous rotary dram chart type thermometer. The U- shaped maximum-minimum thermometer is commonly used for determining diurnal maximum and minimum range of temperature in the culture room.

During the dark period, temperature remains slightly (1-2 °C) lower than light period (all bulbs and tube lights of the culture room increase the temperature). Therefore, chokes of the tube lights are fitted outside the culture room.

The indicators of the maximum-minimum thermometer are moved by mercury column and they remain at that position until moved by the observer with the help of magnet. Their positions indicate the minimum and maximum temperature in the previous 24 h. After recording temperature indicators are reset to mercury level.

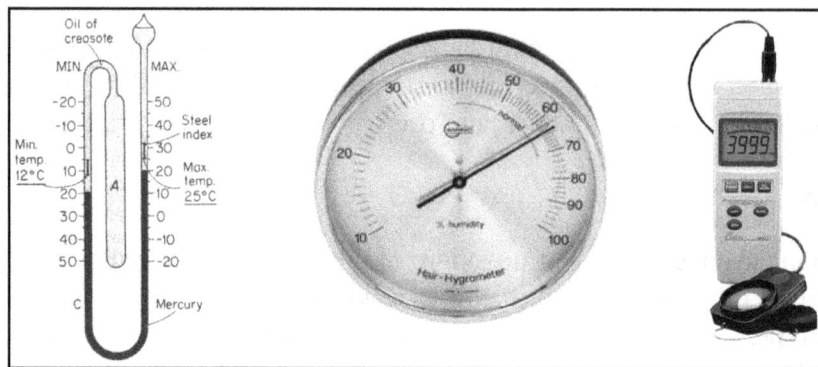

Various accessories required in culture room.

Hygrometer

Hygrometers are instruments used for measuring humidity. A simple form of a hygrometer is specifically known as a "psychrometer" and consists of two thermometers, one of which includes a dry bulb and the other of which includes a bulb that is kept wet to measure wet-bulb temperature. Evaporation from the wet bulb lowers the temperature, so that the wet-bulb thermometer usually shows a lower temperature than that of the dry-bulb thermometer, which measures dry-bulb temperature.

Relative humidity is computed from the ambient temperature as shown by the dry- bulb thermometer and the difference in temperatures as shown by the wet-bulb and dry-bulb thermometers. Relative humidity can also be determined by locating the intersection of the wet- and dry-bulb temperatures on a psychrometric chart.

Dial type hair hygrometer is a convenient tool to measure humidity in the culture room. Humidifier is used to maintain 60% relative humidity (RH) in the culture room. Humidistat is a bimetallic thermocouple device to control and regulate the function of humidifier to maintain the humidity. Distilled water should be filled in the humidifier.

The RH present in the culture room is measured by hair hygrometer. As the name suggests a chemically treated hair elongates with increased humidity and shortens with dryness (similar to mercury in thermometer). It is calibrated from 0 to 100% RH. At lower humidity, medium dries rapidly whereas at higher humidity chances of fungal growth over all surfaces and cotton plugs is increased. Therefore, about 60% RH is maintained in the culture room.

Lux Meter

Light is the form of radiant energy, i.e., electromagnetic radiation of specific wavelength. Visible light as we perceive, is located in narrow wavelength region of spectrum between 380 to 760 nm. Light has dual characters, displaying both wave properties (refraction, diffraction, interference and polarization phenomena) and particle properties (light is radiated in discrete amounts of energy or photons).

Properties of light should be defined either as irradiance (radiant flux intercepted per unit area; unit = w/m²) or an illuminance (luminous flux intercepted per unit area, unit lux, 10.76 lux = 1 foot candle). It is to note that an irradiance measurement is not spectrally defined, whereas, an illuminance measurement indicates the level of visible light as the human eye would see it.

Artificial light is provided by cool, white, fluorescent tubes and/or incandescent bulbs of different classes of artificial light sources, fluorescent sources have been used almost exclusively for plant tissue culture because they are more efficient at producing broad band visible light than incandescent bulbs and are available in low output wattage than lamps.

Light intensity can be measured by photometer or lux meter. A photometer consists of photo-electric cell and a micro-ammeter. Photoelectric cell is sensitive for light and converts light into current. Micro-ammeter shows reading due to this current and its needle moves. Nowadays digital read out is given by appropriately converting the current into digital signal. High intensity is proportional to the current generated in the photoelectric cell by falling light. The unit of illumination is called as 'lux' and the scale of ammeter is calibrated in lux.

Usually 2000 to 2500 lux is provided to the cultures maintained in light in the culture room. The instrument is a sensitive tool and handled with care. It should not be exposed to the sunlight without switching the proper reading switch to high light illumination.

Methods of Sterilization

Asepsis

Plant tissue culture requires contamination free environment, tools and cultures or strict maintenance of germ free system in all the operations, known as asepsis. Particularly in commercial production units, the contamination of one batch of the cultures may result in heavy financial losses or even loss of a culture strain. Therefore strict control measures are enforced to maintain the entry of the personnel and living materials. The basic rules and practices of asepsis are followed in all the tissue culture laboratories.

Plant tissue culture media are rich in nutrients and very suitable for the growth of microbes also. These microorganisms grow faster, consume nutrients rapidly and suppress the growth of plant tissues (by over growth). This saprophytic unwanted growth of microbes is formed as contamination. There are many ways by which these microbes suppress the growth of plant cells and tissues, e.g., release of enzymes and toxins (which inhibit the growth of plant cells), selective absorption of specific nutrients, and some tissue infection.

To achieve success in cultivation of higher plant parts, it is essential to exclude these contaminating microorganisms and hence aseptic techniques must be employed to save cultures. This has led to the development of 'clean area concept'. All the area, tools and working places should be free from microbes to have less and less problems of contamination.

Following care should be taken:

1. Minimize the air current in the working area so it is possible to avoid spores of contaminating microorganisms to move in along with the air currents over the sterile areas. At least, fan should not be used in laminar air flow bench room (inoculation room). Preferably, all the places should be air-conditioned.

2. Store properly the prepared media, nutrients and tools in cabinets.

3. Use separate area for cleaning and washing and for the preparation of medium.

UV tubes are also fixed in Laminar air flow bench placed in inoculation chamber. All these UV tube lights should be used frequently before inoculations. UV lights of corridors may be left open during nights.

Surface Sterilization of Explants

Most contamination is introduced with the explant because of inadequate sterilization or just very dirty material. It can be fungal or bacterial. This kind of contamination can be a very difficult problem when the plant explants material is harvested from the field or greenhouse.

Initial contamination is obvious within a few days after cultures are initiated. Bacteria produce "ooze" on solid medium and turbidity in liquid cultures. Fungi look "furry" on solid medium and often accumulate in little balls in liquid medium.

Bacteria are the most frequent contaminants. They are usually introduced with the explants and may survive surface sterilization of the explants because they are in interior tissues. So, bacterial contamination can first become apparent long after a culture has been initiated. All explants have to be sterilized before transferring them on to the medium.

A suitable sized explants can be sterilized by any one of the following solutions:

1. 1% solution of sodium hypochlorite (commercial bleach).

2. 7 % saturated solution of calcium hypochlorite.

3. 1 % solution of bromine water.

4. 70% ethyl alcohol.

5. 0.2% mercuric chloride.

6. 10% hydrogen peroxide solution.

7. 1% silver nitrate solution.

Suitable sized plant material is sterilized as follows:

1. Clean the working area with ethanol, and start the air flow of the laminar air flow bench.

2. Sterilized petridishes, distilled water, scalpel, filter paper sheets (suitable sized and autoclaved) alcohol, disinfectant (mercuric chloride 0.01-0.1% aqueous solution, or 20% sodium hydrochloride), forceps, bead sterilizer or beaker or coupling jar, sprit lamp or gas burner are collected on Laminar flow bench. Put on the UV light of the bench and put it off after 30 minutes.

3. Bring the plant material to be sterilized on the bench and prepare pieces for sterilization.

4. Clean the working area and hands with alcohol, put on mask and cap, and light the sprit lamp.

5. Keep 3-4 petridishes in a line; add disinfectant in 1st plate and autoclaved distilled water in subsequent plates.

6. Place plant pieces in 1st plate and immerse the material with the help of forceps for 5-10 minutes depending upon the disinfectant used.

7. Transfer material from 1st to 2nd plate, rinse gently and pass to 3rd and 4th plate, one by one with thorough rinsing.

8. Finally, drain the distilled water or place the material in a fresh petridish or filter paper and prepare suitable sized explants.

Mode of Action of Disinfectants

The various metallic ions can be arranged in a series of decreasing antibacterial activity. Hg^{2+} and Ag^+, are effective at less than one part per million (ppm), because of their high affinity for sulphahydryl group. Bacteria are killed by Ag containing 105 to 107 Ag^+ ions per cell. The concentration required for killing is markedly affected by inoculum size.

Chlorine was the anticeptic introduced (as chlorinated lime) by O.W. Holmes in Boston in 1835 and by Semmelweis in Vienna in 1847, to prevent transmission of puerperal sepsis by the physicians hand. Chlorine combines with water to form hypochlorous acid (HOC1) a strong oxidizing agent.

$$CI_2 + H_2O = HCl + HOCl$$
$$Cl_2 + 2NaOH = NaCl + NaOCl + H_2O$$

Sodium hypochlorite (NaOCl) solutions are (200 ppm chlorine) are used to sanitize clean surfaces in the food and the dairy industries and in restaurants. Sodium hypochlorite is commercial bleach available in the market as 3-6% solution and can be used directly. It is prepared by passing chlorine in to dilute solution of sodium hydroxide. Powdered bleach is also added into water supply for killing microbes.

Equipment and Medium Sterilization

All the manipulations of plant tissue culture methods are carried out in aseptic conditions. All the materials used are therefore, free from microbes or sterilized. Various methods of sterilization are used depending upon the material and type of sterilization.

Essentially, method for sterilization is different for living materials than tools and glass ware. Now a day's many items are available as ready to use, pre-sterilized (by gamma radiation) and disposable. Uses of such disposable materials have facilitated the work.

In laboratory, in principle, three types of sterilization is used:

1. Dry heat,
2. Wet heat,
3. Filter sterilization.

Dry Heat

Glassware, metal tools and other articles, which do not get charred by high temperature, are put in containers or wrapped in paper or thick aluminum foil and placed in dry oven and sterilized for a period of not less than three hours at a temperature of 140-160 °C. Media and plastic ware cannot be sterilized by this method.

Wet Heat

The most popular method of sterilization both equipment and media is autoclaving at 121 °C with a pressure of 15 psi (pounds per square inch) for 15 min (1.02 kg/cm²). Modern autoclaves are capable of providing saturated steam treatment ranging from 70-132 °C, which is a pressure of up to 25 psi. All the vessels containing medium should be placed vertically and should not be filled more than 40% to their total capacity.

Filter Sterilization

Filter sterilization or cold sterilization is used when a solution or medium cannot be sterilized by autoclaving. It is the property of the filter (porosity 0.22 to 0.45 μm) to retain the entire microorganism and make the solution free from microbes. This exclusion of microorganisms makes the solution sterilized without heating or autoclaving.

Only liquids can be sterilized by this method and not the plant materials or other things. This is commonly used where thermolabile or heat sensitive chemicals like plant growth regulators (IAA, GA3, Zeatin, Abscisic Acid), some bio-chemicals, enzymes for protoplasts isolation, antibiotics or nutrient solution for specific experiment are used to avoid decomposition during autoclaving. After sterilization, solution is added to the medium. In case of static medium, compound is added before solidification (cooling) after autoclaving.

Glass and membrane filters (nitro-cellulose membranes) are commercially available. Glass (sintered glass filters, Borosil) are available in different grades (G-1 to 5) and the G-5 (1-2 pm) is used for bacteriological purposes. These filters have a porous glass disc for filtration of liquids and gases as a filter media which is non-corrosive and reusable.

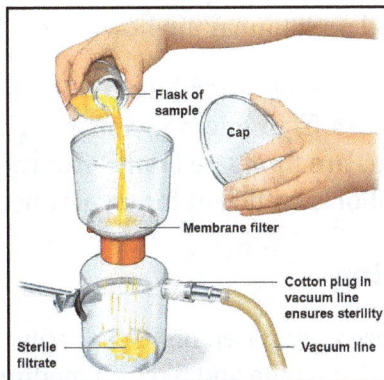

Filtration units of various types.

The grades are classified by maximum pore size which is obtained by measuring the pressure at which the first air bubble breaks away from filter under certain conditions. The pressure differential is then used to calculate the equivalent capillary diameters in microns. The desired pore size is obtained by suitably controlling the grain size, firing time and temperature and the thickness of the disc. These filters are resistant to heat, solvents and detergents.

Membrane type filters (of different diameter) and porosity (0.22 to 0.45 μm) are available as disposable or reusable filters, or as filter holders containing a filter disc. These filters can be connected through tubing or syringe to pass solution or gases. Filters are wrapped in aluminum foil or paper, autoclaved at 121 °C for 15 min and then taken to laminar air flow bench. Solution is placed in funnel (glass type filter) or syringe (membrane type filter) and solution is allowed to pass through filter under suction (glass) or pressure (membrane).

A membrane filter or glass tube, filled with cotton may be placed between vacuum flask and suction pump to avoid entry of air from the outside during operation. Sterilized filtrate is collected in flask and dispensed in the liquid or molten medium with the help of sterilized pipette fitted with cotton plug. Volume added to the medium is adjusted to arrive at a final concentration. Membrane type filters are also used at inlets and outlets of various types including air, in a bioreactor system and to draw samples.

There are a variety of wet and dry heat treatments, radiations, filtration and gas and chemical treatments available for direct sterilization of material. Gas treatments are rarely used in the laboratory. Ultraviolet light treatment of working surfaces and sterile rooms are used in the labs while gamma radiation is used in the industry for the preparation of pre-sterilized disposable plastic-wares.

Sterilization of Tools

Surgical blades and scalpels are not sterilized by dry heat because the high temperature makes the cutting edge dull. Such articles including spatula and forceps are usually immersed in 80% v/v (volume by volume) ethyl alcohol until required, and sterilized during use by frequent immersion in alcohol and flaming. Nowadays, bead sterilizer (works on principle of dry heat) is available for sterilizing such tools.

After an initial stabilization time of 30 minutes during which equipment attains a temperature of about 250 °C, units ensure total sterilization by destruction of all micro-organisms within seconds. This unit can conveniently placed on Laminar air-flow bench.

Uses of fumigation (formaldehyde, sulphur etc.) have been discouraged. It is used some times to clear the laboratory or working area from microorganisms. The place is not used for 3-4 days following fumigation. Culture rooms are fumigated only after removal of cultures. Generally, wiping of working places with 80% alcohol is sufficient for regular use.

Medium and its Preparation

The media used by earlier workers were based on Knop's solution. Subsequently media developed by White and Heller were used. Murashige and Skoog's medium is a land mark in plant tissue culture research and is the most frequently used medium for all types of tissue culture work. Based on its composition, other media were evolved to meet the diverse experimental and species specific requirement.

There are – Linsmaier and Skoog, B5 medium of Gamborg et. al., SH medium of Schenk and Hildebrandt, Nitsch and Nitsch medium, and woody plant medium (WPM) of Llyod and McCown.

The methodology of plant tissue culture has advanced to the stage, where tissues from virtually any plant species can be cultured successfully. The successful plant tissue culture depends upon the choice of nutrient medium. The cells of most plant species can be grown on completely defined media. All the media consist of mineral salts, a carbon source (generally sucrose), vitamins and growth regulators. The MS medium designed for tobacco is now used widely for various species, in callus and cell cultures.

Medium Composition

The nutrient medium for most plant tissue cultures is comprised of five groups of ingredients – inorganic nutrients, carbon source, vitamins, growth regulators and organic supplements.

Inorganic Nutrients

Inorganic nutrients consist of macro- and micro-elements as their salts. Usually nutrient media contain 25 mM each of nitrate and potassium. For regular culture and cell cultures, the combined nitrogen level (nitrate and ammonium nitrogen) may reach up to 60 mM.

Ammonium is essential for most cultures but in lower concentrations than that of nitrate nitrogen. A concentration of 1-3 mM of calcium, magnesium and sulphate, is always adequate. The required micronutrients include I, B, Mn, Zn, Mo, Cu, Co and Fe. Role of different nutrients is given in the table.

Some of the important elements supplied to culture medium for plant nutrition and their physiological function.

Element	Function
Nitrogen	Component of proteins, nucleic acids and some coenzymes, Element required in greatest amount.
Potassium	Regulates osmotic potential, Principal inorganic cation.
Calcium	Cell wall synthesis, member function, cell signalling.
Magnesium	Enzyme cofactor, component of chlorophyll.
Phosphorous	Component of nucleic acids, energy transfer, component of intermediates in respiration and photosynthesis.
Sulphur	Component of some amino acids (methionine, cysteine) and some cofactors.
Chlorine	Required for photosynthesis.
Iron	Electron transfer as a component of cytochromes.
Manganese	Enzyme cofactor.
Cabalt	Component of some vitamins.
Copper	Enzyme cofactor, electron-transfer reactions.
Zinc	Enzyme cofactor, chlorophyll biosynthesis.
Molybdenum	Enzyme cofactor, confactor, component of nitrate reducatse.

Carbon Source

Glucose, fructose, maltose or sucrose (2-4%) can be used as source of energy or carbon but sucrose is the preferred source for most of the cultures. The sucrose in the medium is rapidly converted into glucose and fructose. The glucose is absorbed first followed by fructose.

Vitamins and Amino Acids

Thiamine, pyridoxine and nicotinic acid are commonly used as vitamins in B_5 and MS media. The former is required for most cultures while latter two promote cell growth. Amino acids and organic supplements -Amino acids serve as source of reduced nitrogen. In case of inadequate nitrogen, complex organic nitrogen supplement like casein hydrolysate (0.1-1 g/I) may be supplemented. Glycine is commonly used amino acid. Other organic supplements are coconut milk, yeast extract, peptone and malt extract. However, synthetic media are preferred and organic supplement of unknown chemical nature is used only when it is essential.

Plant Growth Regulators (PGR)

A balanced combination of PGR is required for sustained growth. Two types of combinations are used; one for cell proliferation consisting of (preferably) 2,4- dichlorophenoxy acetic acid (2,4-D) or 1-naphthalene acetic acid (NAA) and a cytokinin (kinetin, benzyl adenosine, 2-isopentyladencisine, zeatin, thidiazuron), another for regeneration essentially containing low auxiti [(NAA, IAA, Indole butyric acid (IBA)] and a cytokinin in high amount, but not 2,4-D as an auxin. 2,4-D is known to induce cell proliferation but suppresses differentiation in dicot plants. However, 2,4-D and 2,4,5-T (2,4,5-trichlorophenoxy acetic acid) are effective in inducing somatic embryogenesis in cereal (monocots) and herbaceous dicot cultures.

Medium Preparation

A convenient approach to prepare a medium is to have stock solutions of all the nutrients in a 10x or 50x concentration. Medium is prepared by suitably diluting the appropriate amount of stock solutions for desired volume of the medium. It may be advantageous to have separate stock solutions of calcium salt and potassium iodide.

All the ingredients are mixed, sugar added and pH is adjusted to 5.8-6.0 and medium is poured in the culture vessels. All the vessels are plugged with non-absorbent cotton, covered with aluminum foil and autoclaved at 121 °C for 15 min.

Prepared media can be stored for a few weeks before inoculation. Liquid medium for a given material is same as static medium used for callus cultures except gelling agent- agar. All the ingredients should be thoroughly mixed before dispensing in the vessels.

Table: Composition of Murashige & Skoog's medium and B_S medium.

Salts	MS		B_5	
A. Macronutrient	Mg/1	mM	Mg/1	mM
NH_4NO_3	1650	20.6	-	-
KNO_3	1900	18.8	2500	25
$CaCi_2.2H_2O$	440	3.0	150	1.0
$MgSO_4.7H_2O$	370	1.5	250	1.0
KH_2PO_4	170	1.25	-	-
$(NH_4)_2SO_4$	-	-	134	1.0
$NaH_2PO_4.H_2O$	-	-	150	1.1
B. Micronutrient	Mg/1	M	Mh/1	μM
KI	0.83	5.0	0.75	4.5
H_3BO_3	6.2	100	3.0	50
$MnSO_4.H_2O$	22.3	100	-	-
$MnSO_4.H_2O$	-	-	10	60
$ZnSO_4.7H_2O$	8.6	30	2.0	7.0
$Na_2MoO_4.2H_2O$	0.25	1.0	0.25	1.0
$CuSO_4.5H_2O$	0.025	0.1	0.025	0.1
$CoCI_2.6H_2O$	0.025	0.1	0.025	0.1
$Na_2.EDTA$	37.3	100	37.3	100
$FeSO_4.7H_2O$	27.8	100	27.8	100
Sucrose (g)	30.000		20,000	
pH	5.7 – 6.0		5.5 – 5.8	
C. Vitamins				
Inositol	100		100	
Nicotinic acid	0.5		1.0	
Pyridoxin HCI	0.5		1.0	
Thiamine. HCI	0.1		10.0	
D. Amino Acid				

Glycine	2.0		-	
E. Plant Growth Regulators				
NAA/IAA/2, 4, -D	0.01-10.0		0.01-10.0	
Kinetin	0.04-1 0.0		0.04-10.0	

Totipotency

Totipotency is the genetic potential of a plant cell to produce the entire plant. In other words, totipotency is the cell characteristic in which the potential for forming all the cell types in the adult organism is retained.

Expression of Totipotency in Culture

The basis of tissue culture is to grow large number of cells in a sterile controlled environment. The cells are obtained from stem, root or other plant parts and are allowed to grow in culture medium containing mineral nutrients, vitamins and hormones to encourage cell division and growth. As a result, the cells in culture will produce an unorganised proliferative mass of cells which is known as callus tissue.

The cells that comprise the callus mass are totipotent. Thus a callus tissue may be in a broader sense totipotent, i.e., it may be able to regenerated back to normal plant given certain manipulations of the medium and the cultural environment. Truly speaking, totipotency of the cell is manifested through the process of differentiation and the hormones in this process play the major role than any other manipulations.

In moving liquid medium some single cells and small groups of cells were loosened from the surface of growing tissue. When these isolated cells were grown separately it was found that some single cells developed somatic embryos or embryoids by a process that occurs in normal zygotic embryo.

It is also observed in some experiment that cells of some callus mass frequently differentiate into vascular elements such as xylem and phloem without forming any plant organs or embryoids. This process is known as histogenesis or Cyto-differentiation. Thus the totipotent cells may express themselves in different way on the basis of differentiation process and manipulation.

Where the totipotent cells are partially expressed or not expressed, it is obvious that the limitation on its capacity for development must be imposed by the microenvironments. The totipotency of cells in the callus tissue may be retained for a longer period through several subcultures.

Practically, it is observed that the ex- plant first forms the callus tissue in the callus inducing medium and such callus tissue is maintained through some subcultures. After then it is generally transferred to another medium which is expected to be favourable for the expression of totipotent cells. Actually, the regeneration medium is standardized by trial and error method.

In more or less suitable medium, the totipotent cells of the callus tissue give rise to meristematic nodules or meristemoids by repeated cell division. This may subsequently give rise to vascular differentiation or it may form a primordium capable of giving rise to a shoot or root. Sometimes the totipotent cell may produce embryoids through sequential stages of development such as globular stage, heart shaped stage and torpedo stage etc.

After prolonged culture, it has been observed that calluses in some species (e.g. Ntcotiana tabacum, Citrus aurantifolia etc.) maybe- come habituated. This means that they are now able to grow on a standard maintenance medium which is devoid of growth hormones. The cells of habituated callus also remain totipotent and are capable to regenerate a plant without any major manipulation.

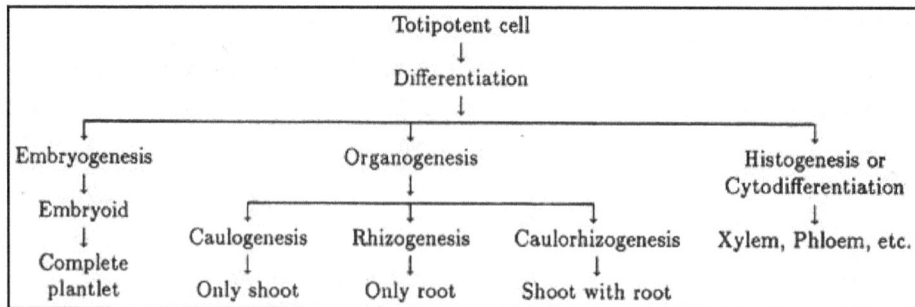

```
                                    Totipotent cell
                                          ↓
                                    Differentiation
                                          ↓
        ┌─────────────────────────────────┼─────────────────────────────────┐
        ↓                                  ↓                                  ↓
   Embryogenesis                     Organogenesis                  Histogenesis or
        ↓                                  ↓                         Cytodifferentiation
   Embryoid           ┌──────────────────┼──────────────────┐               ↓
        ↓             ↓                   ↓                  ↓        Xylem, Phloem, etc.
   Complete      Caulogenesis        Rhizogenesis      Caulorhizogenesis
   plantlet           ↓                   ↓                  ↓
                  Only shoot          Only root        Shoot with root
```

A typical crown gall tumour cell has the capacity for unlimited growth independent of exogenous hormones. It shows totally lack of organ genic differentiation. So such tissue is considered to have permanently lost the totipotentiality of the parent cells.

In some plant species, the crown gall bacterium (Agrobactenum tumefaciens) induces a special type of tumour, called teratomas, the cells of which possess the capacity to differentiate shoot buds and leaves when they are grown in culture for unlimited periods. Thus it is clear that the mode of expression of totipotency of plant cell in culture varies from plant to plant and also helps us to understand the process of differentiation in vitro.

Importance of Totipotency in Plant Science

The ultimate objective in plant protoplast, cell and tissue culture is the reconstruction of plants from the totipotent cell. Although the process of differentiation is still mysterious in general, the expression of totipotent cell in culture has provided a lot of informations.

On the other hand, the totipotentiality of somatic cells has been exploited in vegetative propagation of many economical, medicinal as well as agriculturally important plant species. Therefore, from fundamental to applied aspect of plant biology, cellular totipotency is highly important.

Recent trends of plant tissue culture include genetic modification of plants, production of homozygous diploid plants through haploid cell culture, somatic hybridization, mutation etc. The success of all these studies depends upon the expression of totipotency. In many cases, successful and exciting results have been obtained.

Plant breeders, horticulturists and commercial plant growers are now more interested in plant tissue culture only for the exploitation of totipotent cells in culture according to their desirable requirement. Totipotent cells within a bit of callus tissue can be stored in liquid nitrogen for a long period. Therefore, for germplasm preservation of endangered plant species, totipotency can be utilized successfully.

Organogenesis

Organogenesis is the formation of organs, either shoots or roots. Organogenesis in vitro depends on the balance of auxin and cytokinins and the ability of the tissue to respond to phytohormones during culture. Organogenesis takes place in three phases. In the first phase the cells become competent; next, they differentiate. In the third phase, morphogenesis proceeds independently of the exogenous phytohormones. Organogenesis in vitro can be divided into two types.

Indirect Organogenesis

Formation of organs directly through the callus is called indirect organogenesis. Induction of plants using this technique does not ensure clonal fidelity, but it could be an ideal system for selecting somaclonal variants of desired characteristics and also for mass multiplication. Induction of plants through the callus phase has been used for the production of transgenic plants in which the callus is transformed and the plant regenerated, or the initial explant is transformed and the callus and shoots are developed from the explants.

Direct Organogenesis

The production of direct buds or shoots from a tissue with no intervening callus stage is called direct organogenesis. Plants have been propagated by direct organogenesis for improved multiplication rates, production of transgenic plants, and most importantly for clonal propagation. Typically indirect organogenesis is more important for transgenic plant production. The axillary bud induction/multiple bud initiation technique is the most common means of micropropagation since it ensures the production of uniform planting material without genetic variation. Axillary shoots are formed directly from preformed meristems at nodes, and the chances of the organized shoot meristem undergoing mutation are relatively low. This technique is referred to as multiple bud induction. Many economically important plants have been propagated using this method.

Factors Influencing Organogenesis

Age of Culture

A young culture frequently produces organs. As the culture becomes older, this capacity decreases and ultimately disappears. But there are few exceptions. The; culture of Amorphophallus retains its regeneration capacity indefinitely. Culture of carrot cells can produce roots for many years.

Ploidy Level

In culture there is instability of genome. Only a few weeks after isolation of a diploid callus various degrees of ploidy (mostly tetraploid) have been noticed in various plants (e.g. culture of medullary parenchyma of Nicotiana tabacum, Haplopapptis shoots, pollens of Ginkgo).

Only in a few cases the culture tissues maintain their normal diploid sets, as in culture of tubers of Helianthus tuberosus, leaves of Crepis capillaris and Medicago sativa.

In certain cases, as in culture of pea root callus with increase in ploidy there is decrease in organogenesis. But in some plants, under suitable conditions organs may develop with polyploid meristem. Loss of capacity to organogenesis is reversible in some cases. This may be due to conditions of culture and other non-genetical factors.

Phytohormons

Plant hormones have some effect on organogenesis. auxin at proper concentration can induce root primordia formation in carrot explants. Skoog first said that organogenesis can be chemically controlled. Skoog and Miller observed in tobacco a high auxin cytokinin ratio favours root initiation and a low ratio favours shoot initiation.

Other scientists also observed that under regulated auxin, cytokinin ratio and carbohydrate supply formation of roots and floral or vegetative buds occur. In serveral dicots shoot formation occurs when the ratio between exogenous cytokinin and auxin is 100: 10. But a ratio of 10: 100 favours formation of root primordia.

In monocots shoot initiation occurs on a medium with high 2-4-D and kinetin ratio for four days and then transferred to a medium lacking hormone. In absence of auxin shoot initiation occurs in some cases. For bud initiation certain plants do not require an exogenous supply of cytokinin.

Endogenous gibberellin retards root and shoot initiation. In callus during shoot initiation starch accumulates. Gibberellin lowers the concentration of starch and thereby inhibits shoot initiation.

Endogenous ethylene retards organ initiation during early stages of culture, but in later stages it helps shoot initiation. This has been observed in culture of tobacco cotyledons and Lilium bulb tissue.

According to some scientists phenolic compounds in addition to auxin are more effective for root initiation than auxin alone. On culture of Helianthus tuberosus explants in addition to the auxin the presence of sugar, light, temperature etc. play a major role in root initiation.

Phosphate Concentration

Increase in phosphate concentration favours shoot formation and suppresses or weakens root initiation.

Photoperiodism and Vernalization

Flower formation on culture can be induced by photoperiodism in Plumbago indica and by vernalization in Cichorium intybus and Lunaria annua.

Plantlet Formation form Tomato Leaves

From surface sterilised young leaves rectangular explants (6 mm x 8 mm) are cut out. Each explants is placed on a tube containing 15 c.c. of Murashige and Skoog's medium, which is composed of mineral salts, thiamine (0.4 mg/l), myo-inositol (100 mg/l), sucrose (30,000 mg/l) and a growth regulator.

For root initiation 2 mg/l IAA and 2 mg/l kinetin are needed at 12 hours photoperiod and a temperature of 25 °C. After initiation of root the callus is transferred to a medium containing 4 mg/l kinetin and 4 mg/l IAA. In this medium shoot "initiation occurs after four weeks. Culture is transferred to a medium without any hormone. Subculture can be done every 3-4 weeks.

Artificial Seeds

Artificial seeds are most commonly described as encapsulated somatic embryos. They are product of somatic cells, so can be used for large scale clonal propagation. Apart from somatic embryos, other explants such as shoot tips, axillary buds have also been used in preparation of artificial seeds. Artificial seeds have a variety of applications in plant biotechnology such as large scale clonal propagation, germplasm conservation, breeding of plants in which propagation trough normal seeds is not possible, genetic uniformity, easy storage and transportation etc. For some plants such as ornamental plants, propagation trough somatic embryogenesis and artificial seeds is the only way out.

The seed (or zygotic seed) is the vehicle that connects one generation to another in much of the plant kingdom. By means of seed, plants are able to transmit their genetic constitution in generations and therefore seeds are the most appropriate means of propagation, storage and dispersal. Artificial seeds have great potential for large scale production of plants at low cost as an alternative to true seeds. An artificial seed is often described as a novel analogue to true seed consisting of a somatic embryo surrounded by an artificial coat which is at most equivalent to an immature zygotic embryo, possibly at post-heart stage or early cotyledonary stage. There are various advantages of artificial seeds such as; better and clonal plants could be propagated similar to seeds; preservation of rare plant species extending biodiversity could be realized; and more consistent and synchronized harvesting of important agricultural crops would become a reality, among many other possibilities. In addition; ease of handling, potential long-term storage and low cost of production and subsequent propagation are other benefits.

The artificial seed production technique was first used in clonal propagation to cultivate somatic embryos placed into an artificial endosperm and constrained by an artificial seed coat. Today artificial seeds represent capsules with a gel envelope, which contain not only somatic embryos but also axillary and apical buds or stem and root segments. Explants such as shoot tips, axillary buds and somatic embryos are encapsulated in cryoprotectant material like hydrogel, alginate gel, ethylene glycol, dimethylsulfoxide (DMSO) and others that can be developed into a plant. The coating protects the explants from mechanical damage during handling and allows germination and conversion to occur without inducing undesirable variations. They behave like true seeds and sprout into seedlings under suitable conditions.

The Need for Artificial Seed

A seed is basically zygotic embryo with enhanced nutritive tissues and covered by several protective layers. Seeds are desiccation tolerant, durable and quiescent due to protective coat. Such properties of seeds are also used for germplasm preservation in seed repositories. Zygotic embryo seeds are the result of sexual reproduction that means the progeny of two parents. This has led to the development of often complex breeding programs from which inbred parental lines are developed. Such inbred lines are used to produce uniform hybrid progeny when crossed. Primary problem associated with such seeds is, on one hand for many crops, such as fruits, nuts, and certain ornamental plants; it is not possible to produce a true-breeding seed from two parents due to genetic barriers to selfing. On the other hand many crops, such as forest trees, the generation time is too long to achieve rationally an inbred breeding program. This is the major disadvantage of zygotic seeds. Therefore, for such crops, propagation is accomplished either vegetatively by cuttings or the use of relatively lowquality open pollinated seed is tolerated.

Somatic embryo arises from the somatic cells of a single parent. They differ from zygotic embryos since somatic embryos are produced through in vitro culture, without nutritive and protective seed coats and do not typically become quiescent. Somatic embryos are structurally equivalent to zygotic embryos, but are true clones, since they arise from the somatic cells of a single parent. The structural complexity of artificial seeds depends on requirements of the specific crop application. Therefore, a functional artificial seed may or may not require a synthetic seed coat, be hydrated or dehydrated, quiescent or non quiescent, depending on its usage. The field that seeks to use somatic embryos as functional seed is termed "artificial or synthetic seed technology". Thus, artificial seeds are defined from a practical standpoint as somatic embryos engineered to be of use in commercial plant production and germplasm preservation.

Types of Artificial Seeds

There are various types of artificial seeds; first two are essentially uncoated somatic embryos; (i) uncoated non quiescent somatic embryos, which could be used to produce those crops that are now laboriously micro propagated by tissue culture; (ii) uncoated, quiescent somatic embryos would be useful for germplasm storage since they can be hand-stored in existing seed storage repositories. The other categories are; (iii) Non quiescent somatic embryos in a hydrated encapsulation constitute a type of artificial seed that may be cost effective for certain field crops that pass through a greenhouse transplant stage such as carrot, celery, seedless watermelon, and other vegetables and iv. dehydrated, quiescent somatic embryos encapsulated in artificial coatings are the form of artificial seed that most resembles conventional seed in storage and handling qualities. These consist of somatic embryos encased in artificial seed coat material, which then is dehydrated. Under these conditions, the somatic embryos become quiescent and the coating hardens. Theoretically, such artificial seeds are durable under common seed storage and handling conditions. Upon rehydration, the seed coat softens, allowing the somatic embryo to resume growth, enlarging and emerging from the encapsulation. Many studies have been conducted on synthetic seed production in horticultural crops but the efforts in field grown crops are limited. So, there is a greater scope for synthetic seeds in commercial crops and ornamental plants.

Advantages of Artificial Seeds

There are various advantages of artificial seeds. One of the chief advantages is the possibility of large scale propagation and mixed genotype plantations – very much suitable for large scale monoculture. Another big advantage is the germplasm conservation of elite and endangered or extinct plant species. Other advantages are easy handling during storage; transportation and planting and inexpensive transport reason being their small size; storage life comparable to natural seeds; product uniformity – as somatic embryos used are genetically identical. In addition, other potential benefits can be direct field use, study of seed coat formation, fusion of endosperm in embryo development and seed germination; for production of hybrids in plants with unstable genotypes or show seed sterility. It can be used in combination with embryo rescue technique.

Procedure for the Production of Artificial Seeds

There could be a number of possible artificial seed systems, depending upon the type of artificial seed produced, need of artificial seeds, the economic feasibility and it will vary greatly among species. The development of the ideal viable, quiescent, low-cost artificial seed has been described as a 10-step process. First of the steps is the selection of the crop based on technological and commercial potential followed by the establishment of a somatic embryo system (speciesspecific). Next is the optimization of the clonal production system (optimizing protocols to synchronize and maximize the development of normal mature embryos capable of conversion to normal plants. Automation of embryo production is followed by this. After that, post-treatment of mature embryos to induce quiescence, development of an encapsulation and coating system, optimization and automation of the encapsulation system and conversion requirements for greenhouse and field growth (watering, fertilizer, transplantation, etc.) are followed. Identification and control of any pest and disease problems that may be unique to artificial seeds and determination of the economic feasibility of using the artificial seed delivery system for a specific crop compared with other propagation methods (cost–benefit analysis of encapsulation versus other options) are last steps. Some steps generally apply to more than one species whereas other steps may be species-specific. The latter are inevitably the most demanding in terms of development, and are noted as such.

Applications of Artificial seeds

Artificial seeds have vast application in different fields of plant biotechnology for cultivation of various plant species. They offer the opportunity to store genetically heterozygous plants or plants with a single outstanding combination of genes that could not be maintained by conventional methods of seed production due to genetic recombination exists in every generation for seed multiplication.

Many species are sterile and produce no seeds. Somatic embryogenesis is an alternative with respect to the cuttings to propagate these plants. Other species, including some tropical produce recalcitrant seeds that can not be dried. Consequently, long-term storage in gene banks in these species is not possible. The artificial seeds can be an alternative as more is learned about the mechanism by which this type of seed has no tolerance to desiccation. In autogamous species, where the production of hybrid seed is difficult and expensive, the artificial seed technology offers many advantages and opportunities.

One of the limitations of the method of micro propagation is that they should be in the same physical site of tissue culture laboratories and greenhouses, as production of propagules must be synchronized in periods of peak demand in the market. Artificial seed production in these species would not link the laboratory facilities of the greenhouses.

The market for ornamental plants is growing every year. The high cost of production of these species is given by the diligence of the micro propagation and manpower needed in the later stages of propagation and production. The use of somatic embryogenesis system in these species would significantly reduce labor costs.

Coniferous forest species can be propagated cheaply through seeds. The conventional breeding programs in these species are very time consuming because the life cycle of conifers is very long. Coniferous forests are very heterogeneous and that the seed of outstanding individuals will not necessarily give rise to improved offspring. Artificial seed has the ability to clone those overhanging trees at reasonable cost and in minimum time.

In the commercial sector, it is very difficult to produce low-cost hybrid seed species such as cotton (Gossypium hirsitum L.) and soybean (Glycine max Merril.) because they have cleistogamous flowers and abscission problems as the seed that is currently used comes from self-pollinating species. However, hybrid seed is produced in small quantities in a very laborious by hand pollination. This small volume of hybrid seed could be massively increased through artificial seed technology. Thus, the hybrid force would be used commercially to originate a significant reduction in costs.

In certain vegetable species, used hybrid seeds are expensive and therefore the plant value is very high. For example, tomatoes and seedless watermelon hybrid seeds are used in very high cost. The reason for this high cost is that pollination is done by hand, requiring intensive labor. In other species, vegetative reproduction is used it also consumes much time, space and labor. The use of artificial seed technology can significantly reduce costs by reducing the labor required, time and space in case of these plants.

Sowing seed of synthetic varieties is a common practice in forage species such as alfalfa (Medicago sativa L.) and orchargrass (Dactylis glomerata L.). Such varieties from selection and crossing of lines are phenotypically uniform but different genotypes. These lines to cross freely year after year to produce seeds, heterozygous and heterogeneous populations originate. The use of artificial seed allows multiplication of outstanding genotypes and genetically uniform, since this method does not require that annually cross-pollination is carried out to produce plants.

The vast majority of fruit species are propagated by vegetative means because of the presence of self-incompatibility and breeding cycles very long. The use of synthetic seed facilitates its spread. However, the most useful artificial seed would be in the conservation of germplasm of these species. Currently seed banks are maintained as live plants in the field. This method of conservation is very expensive and dangerous, as it is exposed to natural disasters. The use of artificial seeds would retain these clones in a small space, under controlled conditions (cryopreservation) and without the danger of natural disasters. In addition, this system of germplasm conservation would be particularly useful in tropical species where conservation means are inadequate or nonexistent. The vine (Vitis spp.) is a practical example of this system of conservation.

In cross-pollinated species like maize, where production of hybrid seed is a widespread practice. The creation of hybrids through a conventional breeding program consumes much time and resources in obtaining and maintaining appropriate parental lines. One possibility is the use of artificial seed to propagate outstanding genotypes without the need to generate parental lines costly in time and money. This could facilitate the commercialization of new hybrids and encourage the emergence of new seed companies, as it would be possible to produce new hybrids without the need for large amounts of parental lines.

In autogamous species such as wheat, barley and oats where hybrid seed production at commercial level is not possible by high production costs, artificial seeds would spread the hybrid seed. In this case, produce small quantities of hybrid seed by hand and then with the technology of artificial seed multiplication would be carried out mass.

There are a growing number of species that are in the process of extinction. Indiscriminate felling of forests, increasing desertification, disappearing forests, etc. increases the changes of extinction of species. Many of these native species cannot be propagated vegetatively, or produce very low quantities of seed. For this reason the artificial seed is an alternative for these species. For example, in Australia, eucalyptus tolerant to saline soils has been obtained. These eucalypts cannot multiply by cuttings or by seed true. One option is the artificial seed technology.

Crops from genetically modified plants have boomed in recent years. There is little information about what happens to these GMOs in the process of sexual reproduction. It is possible that during sexual multiplication, the introduced genes from other species are meiotically unstable and lost. With the use of artificial seed technology would avoid such risks. Similarly, this technology could be used in the propagation of somatic hybrids and cytoplasmic (obtained through protoplast fusion) and in sterile and unstable genotypes.

Artificial seeds of Dendrocalamus strictus, commonly called the male bamboo, were produced by encapsulating somatic embryos that had been obtained on MS medium containing 3.0 mg 1 −I 2,4- dichlorophenoxyacetic acid (2,4-D) and 0.5 mg I -I kinetin (Kin), in calcium alginate beads. A germination frequency of 96% and 45% was achieved in vitro and in soil, respectively. The in vivo plantlet conversion frequency was increased to 56% following an additional coating of mineral oil on the alginate beads. They were able to achieve the germination of artificial seeds into plantlets.

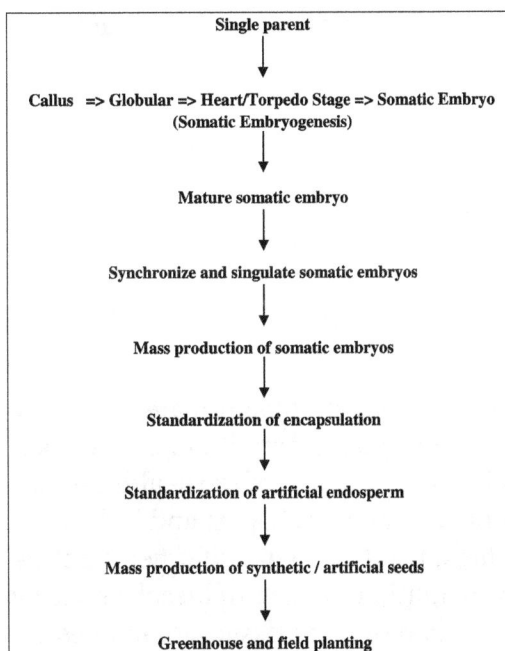

Procedure for production of artificial seed.

Micropropagation

Micropropagation is the practice of rapidly multiplying stock plant material to produce a large number of progeny plants, using modern plant tissue culture methods.

A rose plant that began as cells grown in a tissue culture.

Micropropagation is used to multiply plants such as those that have been genetically modified or bred through conventional plant breeding methods. It is also used to provide a sufficient number of plantlets for planting from a stock plant which does not produce seeds, or does not respond well to vegetative reproduction. Cornell University botanist Frederick Campion Steward discovered and pioneered micropropagation and plant tissue culture in the late 1950s and early 1960s.

Steps

In short, steps of micropropagation can be divided into 4 stages:

1. Selection of mother plant,
2. Multiplication,
3. Rooting and acclimatizing,
4. Transfer new plant to soil.

Establishment

Micropropagation begins with the selection of plant material to be propagated. The plant tissues are removed from an intact plant in a sterile condition. Clean stock materials that are free of viruses and fungi are important in the production of the healthiest plants. Once the plant material is chosen for culture, the collection of explant(s) begins and is dependent on the type of tissue to be used; including stem tips, anthers, petals, pollen and other plant tissues. The explant material is then surface sterilized, usually in multiple courses of bleach and alcohol washes, and finally rinsed in sterilized water. This small portion of plant tissue, sometimes only a single cell, is placed on a growth medium, typically containing sucrose as an energy source and one or more plant growth regulators (plant hormones). Usually the medium is thickened with agar to create a gel which supports the explant during growth. Some plants are easily grown on simple media, but others require more complicated media for successful growth; the plant tissue grows and differentiates into new tissues depending on the medium. For example, media containing cytokinin are used to create branched shoots from plant buds.

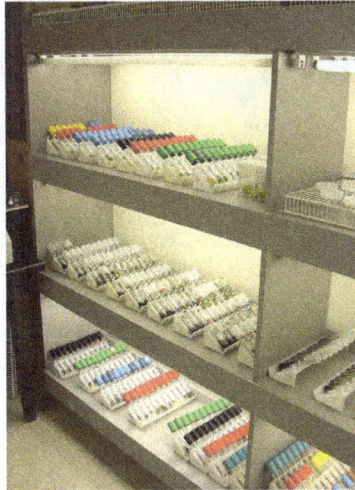

In vitro culture of plants in a controlled, sterile environment.

Multiplication

Multiplication is the taking of tissue samples produced during the first stage and increasing their number. Following the successful introduction and growth of plant tissue, the establishment stage is followed by multiplication. Through repeated cycles of this process, a single explant sample may be increased from one to hundreds and thousands of plants. Depending on the type of tissue grown, multiplication can involve different methods and media. If the plant material grown is callus tissue, it can be placed in a blender and cut into smaller pieces and recultured on the same type of culture medium to grow more callus tissue. If the tissue is grown as small plants called plantlets, hormones are often added that cause the plantlets to produce many small offshoots. After the formation of multiple shoots, these shoots are transferred to rooting medium with a high auxin\cytokinin ratio. After the development of roots, plantlets can be used for hardening.

Pretransplant

This stage involves treating the plantlets/shoots produced to encourage root growth and "hardening." It is performed in vitro, or in a sterile "test tube" environment.

"Hardening" refers to the preparation of the plants for a natural growth environment. Until this stage, the plantlets have been grown in "ideal" conditions, designed to encourage rapid growth. Due to the controlled nature of their maturation, the plantlets often do not have fully functional dermal coverings. This causes them to be highly susceptible to disease and inefficient in their use of water and energy. In vitro conditions are high in humidity, and plants grown under these conditions often do not form a working cuticle and stomata that keep the plant from drying out. When taken out of culture, the plantlets need time to adjust to more natural environmental conditions. Hardening typically involves slowly weaning the plantlets from a high-humidity, low light, warm environment to what would be considered a normal growth environment for the species in question.

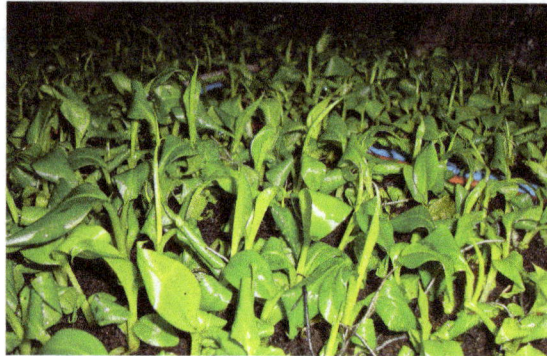

Banana plantlets transferred to soil (with vermicompost) from plant media.
This process is done for acclimatization of plantlets to the soil as they were previously grown in plant media. After growing for some days the plantlets are transferred to the field.

Transfer from Culture

In the final stage of plant micropropagation, the plantlets are removed from the plant media and transferred to soil or (more commonly) potting compost for continued growth by conventional methods. This stage is often combined with the "pretransplant" stage.

Plant tissue cultures being grown at a USDA seed bank,
the National Center for Genetic Resources Preservation.

Methods

Meristem Culture

In Meristem culture the Meristem and a few subtending leaf primordial are placed into a suitable growing media. An elongated rooted plantlet is produced after some weeks, and is transferred to the soil when it has attained a considerable height. A disease free plant can be produced by this method. Experimental result also suggest that this technique can be successfully utilized for rapid multiplication of various plant materials, e.g. Sugarcane, strawberry

Callus Culture

A callus is mass of undifferentiated parenchymatous cells. When a living plant tissue is placed in an artificial growing medium with other conditions favorable, callus is formed. The growth of callus varies with the homogenous levels of auxin and Cytokinin and can be manipulated by endogenous supply of these growth regulators in the culture medium. The callus growth and its organogenesis or embryogenesis can be referred into three different stages:

- Stage I: Rapid production of callus after placing the explants in culture medium.

- Stage II: The callus is transferred to other medium containing growth regulators for the induction of adventitious organs.

- Stage III: The new plantlet is then exposed gradually to the environmental condition.

Suspension Culture

A cell suspension culture refers to cells and or groups of cells dispersed and growing in an aerated liquid culture medium is placed in a liquid medium and shaken vigorously and balanced dose of hormones. Suezawa et al. reported Cyotkinin induced adventitious buds in kiwi fruit in a suspension culture sub- culture for about a week.

Embryo Culture

In embryo culture, the embryo is excised and placed into a culture medium with proper nutrient in aseptic condition. To obtain a quick and optimum growth into plantlets, it is transferred to soil. It is particularly important for the production of interspecific and intergeneric hybrids and to overcome the embryo.

Protoplast Culture

In protoplast culture, the plant cell can be isolated with the help of wall degrading enzymes and growth in a suitable culture medium in a controlled condition for regeneration of plantlets. Under suitable conditions the protoplast develops a cell wall followed by an increase in cell division and differentiation and grows into a new plant. The protoplast are first cultured in liquid medium at 25 to 28 C with a light intensity of 100 to 500 lux or in dark and after undergoing substantial cell division, they are transferred into solid medium congenial or morphogenesis in many horticultural crops respond well to protoplast culture.

Advantages

Micropropagation has a number of advantages over traditional plant propagation techniques:

- The main advantage of micropropagation is the production of many plants that are clones of each other.

- Micropropagation can be used to produce disease-free plants.

- It can have an extraordinarily high fecundity rate, producing thousands of propagules while conventional techniques might only produce a fraction of this number.

- It is the only viable method of regenerating genetically modified cells or cells after protoplast fusion.

- It is useful in multiplying plants which produce seeds in uneconomical amounts, or when plants are sterile and do not produce viable seeds or when seed cannot be stored.

- Micropropagation often produces more robust plants, leading to accelerated growth compared to similar plants produced by conventional methods - like seeds or cuttings.

- Some plants with very small seeds, including most orchids, are most reliably grown from seed in sterile culture.

- A greater number of plants can be produced per square meter and the propagules can be stored longer and in a smaller area.

Disadvantages

Micropropagation is not always the perfect means of multiplying plants. Conditions that limits its use include:

- It is very expensive, and can have a labour cost of more than 70%.

- A monoculture is produced after micropropagation, leading to a lack of overall disease resilience, as all progeny plants may be vulnerable to the same infections.

- An infected plant sample can produce infected progeny. This is uncommon as the stock plants are carefully screened and vetted to prevent culturing plants infected with virus or fungus.

- Not all plants can be successfully tissue cultured, often because the proper medium for growth is not known or the plants produce secondary metabolic chemicals that stunt or kill the explant.

- Sometimes plants or cultivars do not come true to type after being tissue cultured. This is often dependent on the type of explant material utilized during the initiation phase or the result of the age of the cell or propagule line.

- Some plants are very difficult to disinfect of fungal organism.

The major limitation in the use of micropropagation for many plants is the cost of production; for many plants the use of seeds, which are normally disease free and produced in good numbers, readily produce plants in good numbers at a lower cost. For this reason, many plant breeders do not utilize micropropagation because the cost is prohibitive. Other breeders use it to produce stock plants that are then used for seed multiplication.

Mechanisation of the process could reduce labour costs, but has proven difficult to achieve, despite active attempts to develop technological solutions.

Somatic Embryogenesis

In plant tissue culture, the developmental pathway of numerous well-organised, small embryoids resembling the zygotic embryos from the embryo genic potential somatic plant cell of the callus tissue or cells of suspension culture is known as somatic embryogenesis.

Embryo Genic Potential

The capability of the somatic plant cell of a culture to produce embryoids is known as embryo genic potential.

Embryoid

Embryoid is a small, well-organised structure comparable to the sexual embryo, which is produced in tissue culture of dividing embryo genic potential somatic cells.

Principles of Somatic Embryogenesis

Somatic embryogenesis may be initiated in two different ways:

1. In some cultures somatic embryogenesis occurs directly in absence of any callus production from "pro-embryo genic determined cells" that are already programmed for embryo differentiation. For instance, somatic embryos has been developed directly from leaf mesophyll cells of orchard grass (Dactyhs glomerata L.) without an intervening callus tissue.

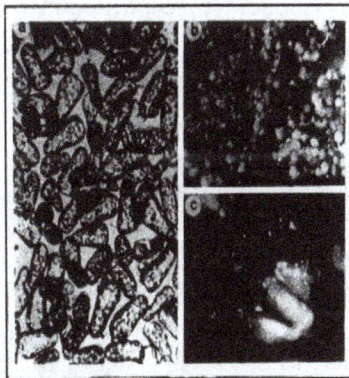

Photograph showing direct embryogenesis. A. A suspension of Mechanically isolated mesophyll cells. B. Embryogenesis. C. A portion enlarged from B.

Explants, made from the basal portions of two innermost leaves of orchard grass were cultured on a Schenk and Hildebrandt medium supplemented with 30 μM 3, 6-dichloro-O-anisic acid (dicamba). Plant formation occurred after sub culturing the embryos on the same medium without dicamba.

> **Routes of Somatic Embryogenesis:-**
> Two routes to somatic embryogenesis
>
> ❖ Direct somatic embryogenesis
> ➢ The embryos initiate directly from explants in the absence of callus formation. Embryos are formed due to PEDCs cell.
>
> ❖ Indirect somatic embryogenesis
> ➢ Callus from explants takes place from which embryos are developed. Embryos are formed due to IEDCs cells.

2. The second type of somatic embryo development needs some prior callus formation and embryoids originate from "induced embryo genic cells" within the callus tissue.

In most of the cases, indirect embryogenesis occurs. For indirect somatic embryogenesis where it has been induced under in vitro condition, two distinctly different types of media may be required—One medium for the initiation, of the embryonic cells and another for the subsequent development of these cells into embryoids.

The first or induction medium must contain auxin in case of carrot tissue and somatic embryogenesis can be initiated in the second medium by removing the hormone or lowering its concentration. With some plants, however, both embryo initiation and subsequent maturation and subsequent maturation occur on the first medium and a second medium is required for plantlet development.

In some cases, a given culture may differentiate the embryo genic cells, but their further growth may be blocked by an imbalance of nutrition in the culture medium. According to Kohlenback, abnormalities known as embryonal budding and embryo genic clump formation may occur, if relatively high level of auxin is present after the embryo genic cells have been differentiated.

Embryoids are generally initiated in callus tissue from the superficial clumps of cells (primordia) associated with enlarged vacuolated cells that do not take part in embryogenesis. The embryo genic cells are generally characterised by dense cytoplasmic contents, large starch grains, a relatively large nucleus with a darkly stained nucleolus. In suspension culture, embryoids do not form suspended single cell, but form cells lying at or near the surface of the small cell aggregates.

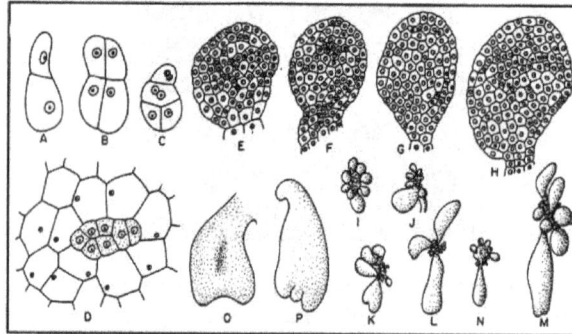

Stages in development of embryolds within Atropa suspension aggregates.

A-C. Early stages in the development of a proembryold. D. Part of a cell aggregate showing an embedded proembryold. E.-H. Progressive stages in development of a globular embryold. I.-N. Cell aggregates from suspension culture showing various stages of embryongeny. O and P. Later stages in embryogeny.

Each developing embryoid of carrot passes through three sequential stages of embryo formation such as globular stage, heart-shape stage and torpedo stage. The torpedo stage is a bipolar structure which ultimately gives rise to complete plantlet. The culture of other plants may not follow such sequential stages of embryo development.

In general, somatic embryogenesis occurs in short-term culture and this ability decreases with increasing duration of culture. But there are some exceptional cultures where embryogenesis has been reported from the callus tissue maintained over a period of year. The loss of embryo genic potential in long term culture may also result from loss of certain biochemical properties of the cell.

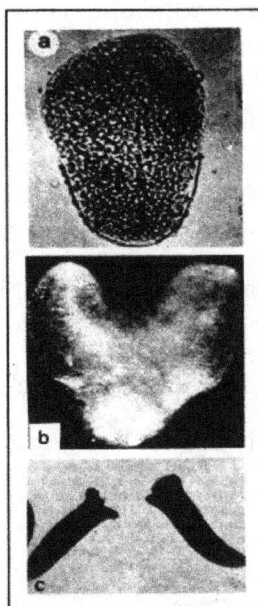

Embryoid development In tissue culture passes through various stages, namely
(a) globular stage (b) heart shaped stage and (c) torpedo stage before plantlet formation.

In callus culture or in suspension culture, embryoid formation occurs asynchronously. Some progress has been made in inducing synchronization of somatic embryogenesis in cell suspension culture. A high degree of synchronization has been achieved in a carrot suspension culture by sieving the initial cell population.

Protocols for Inducing Somatic Embryogenesis in Culture

The plant material Daucus carota represents the classical example of somatic embryogenesis in culture.

The protocol is described below:

1. Leaf petiole (0.5-1 cm) or root segments from seven-day old seedlings (1 cm) or cambium tissue (0.5 cm3) from storage root can be used as explant. Leaf petiole and root segment can be obtained from aseptically grown seedlings (Cambium tissue can be obtained from surface sterilized storage tap root.

2. Following aseptic technique, explants are placed individually on a semi-solid Murashige and Skoog's medium containing 0.1 mg/L 2, 4-D and 2% sucrose. Cultures are incubated in the dark. In this medium the explant will produce sufficient callus tissue.

3. After 4 weeks of callus growth, cell suspension culture is to be initiated by transferring 0.2 gm. of callus tissue to a 250 ml of Erlenmeyer flask containing 20-25 ml of liquid medium of the same composition as used for callus growth (without agar). Flasks are placed on a horizontal gyratory shaker with 125-160 rpm at 25 °C. The presence or absence of light is not critical at this stage.

4. Cell suspensions are sub-cultured every 4 weeks by transferring 5 ml to 65 ml of fresh liquid medium.

5. To induce a more uniform embryo population, cell suspension is passed through a series of stainless steel mesh sieves. For carrot, the 74 µ sieve produces a fairly dense suspension of single cell and small multiple clumps. To induce somatic embryogenesis, portions of sieved cell suspension are transferred to 2, 4-D free liquid medium or cell suspension can be planted in semi-solid MS medium devoid of 2, 4-D. For normal embryo development and to inhibit precocious germination especially root elongation, 0.1- 1 µM ABA can be added to the culture medium. Cultures are incubated in dark.

6. After 3-4 weeks, the culture would contain numerous embryos in different stages of development.

7. Somatic embryos can be placed on agar medium devoid of 2, 4-D for plantlet development.

8. Plantlets are finally transferred to Jiffy pots or vermiculite for subsequent development.

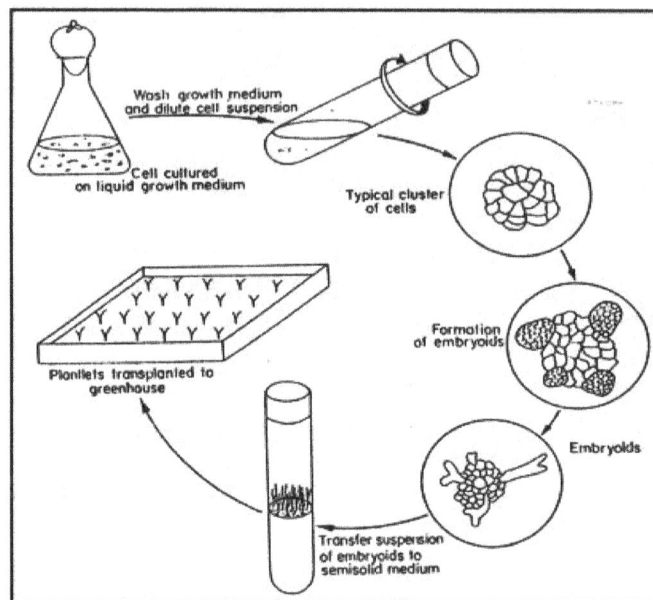

Flow diagram illustrating the protocol for inducing somatic embryogenesis in culture.

Importance of Somatic Embryogenesis

The potential applications and importance of in vitro somatic embryogenesis and organogenesis are more or less similar. The mass production of adventitious embryos in cell culture is still regarded by many as the ideal propagation system. The adventitious embryo is a bipolar structure that develops directly into a complete plantlet and there is no need for a separate rooting phase as with shoot culture.

Somatic embryo has no food reserves, but suitable nutrients could be packaged by coating or encapsulation to form some kind of artificial seeds. Such artificial seeds produce the plantlets directly into the field. Unlike organogenesis, somatic embryos may arise from single cells and so it is of special significance in mutagenic studies.

Plants derived from asexual embryos may in some cases be free of viral and other pathogens. For an example, Citrus plant propagation from embryo genic callus of nuclear origin are free of Virus. So it is an alternative approach for the production of disease-free plants.

Meristem and Shoot Tip Culture

Meristem culture particularly involves the cultivation of the shoot apical meristem. This technique also refers to Shoot tip or Apical meristem culture.

The meristem can culture by isolating from the stem by applying a V-Shape cut. In this technique, by the culturing of shoot meristem, adventitious roots can regenerate. The size of shoot tip is 10mm for the generation of the virus-free plant, and for vegetative propagation, the size of the shoot tip does not matter.

The technique of meristem culture is very much similar to the technique of micropropagation which also involves:

- Initiation of culture,
- Multiplication of culture,
- Rooting of a developed shoot,
- Transfer of plantlets to the pots.

For the regeneration of plant by the meristem culture, the most widely used medium is Murashige and Skoog's (MS) medium with low salt concentration for the majority of the species. Fungicides (Bavistin) or antibiotics (chloramphenicol or streptomycin) uses to remove the endophytic contamination.

Meristem culture can define as the tissue culture technique which makes the use of apical meristem with 1-3 leaf primordia by which clones of a plant can develop by the vegetative propagation.

Meristem

Meristem is the part of a plant which plays a key function to increase the plant length. It composes of cells refers to a meristematic cell which are the continuous cells that are oval, polygonal, rectangle in shape. There is no intercellular space between the meristematic cells.

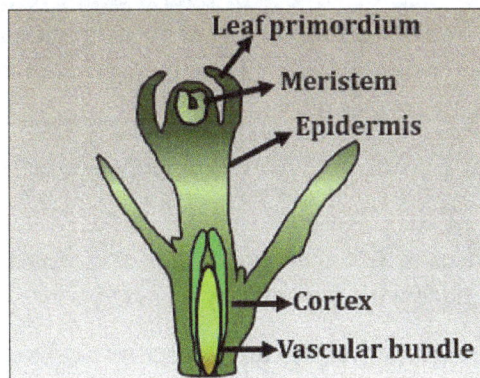

Meristem is the shoot tip or the apical region which is found at the shoot apical and root apical region. It is "Dome-shaped" whose thickness is about 0.1 mm in diameter and 0.25-0.3 mm in length.

The apical shoot meristem is the portion where a stem elongates and is free of pathogens which were proved in 1949, before the introduction of meristem culture. Limmaset and Cornuet were the two scientists who observed the declined growth of virus towards the apical meristem. For this reason, meristem culture is very popular among other techniques in the production of virus and other related organisms free plants.

Explant Terminology

The apical meristem of a shoot is the portion lying distal to the youngest leaf primordium, it measures up to about 100µm in diameter and 250µm in length. The apical meristem together with one to three young leaf primordia, measuring 100-500µm, constitutes the shoot-apex. Although the chances of eradicating viruses is higher through 'meristem' culture, in most successful reports virus-free plants have been raised by culturing 100-1000µm long explants which could be according to the above definition is referred as 'shoot-tip'. To distinguish it from the in vivo technique of propagation through shoot-tip cuttings, the term 'meristem-tip' culture has been preferred for in vitro culture of small shoot-tips.

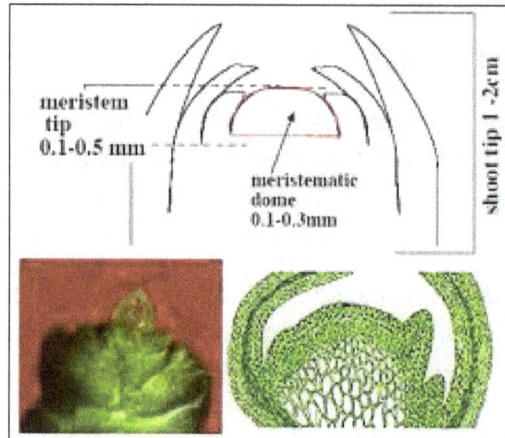

Reasons for Escape of Meristem from Virus

It is well known that the distribution of viruses in plants is uneven. In infected plants the apical meristems are generally either free or carry a very low concentration of the viruses. In older tissues the virus titer increases with increasing distance from the meristem-tips. The reasons proposed for the escape of meristem from virus invasion are:

a) Viruses readily move in a plant body through the vascular system which is absent in the meristem.

b) The alternative method of cell-to-cell movement of the virus through plasmodesmata is rather too slow to keep pace with the actively growing tip.

c) High metabolic activity in the actively dividing meristem cells does not allow virus replication.

d) The 'virus inactivating systems' in the plant body, if any, has higher activity in the meristem than in any other region. Thus, the meristem is protected from infection.

e) A high endogenous auxin level in shoot apices may inhibit virus multiplication.

Factors Affecting Eradication of Virus through Meristem Tip Culture

Culture medium, explant size and incubation conditions affecting plant regeneration from meristem-tip cultures have pronounced effect on virus eradication. Besides, thermotherapy or chemotherapy and physiological stage of the explants also affect virus elimination by shoot-tip culture.

Culture Medium

The nutrients, growth regulators and nature of the medium highly influence the development of virus free plants from meristem tip cultures. Maximum success is achieved from Murashige & Skoog's (MS) medium which promoted healthy, green shoot development compare to other nutrient media. The main reason for the suitability of medium for meristem-tip culture could be the presence of high levels of K^+ and NH_4^+ ions. There is no critical assessment on the role of various vitamins or amino acids but sucrose or glucose is the most commonly used carbon source in the medium, at the range of 2-4%, to raise virus free plants from meristem-tip cultures.

Large meristem-tip explants, measuring 500µm or more in length, may give rise to plants even in the basal medium but generally the presence of an auxin or a cytokinin or both plays a major role in the development of excised apical meristem. In angiosperms, the meristematic dome in the shoot-tip does not synthesize auxin on its own, but it is supplied by the second pair of youngest leaf primordia. Therefore, for development of excised meristem in culture, without the leaf primordia, requires the supply of exogenous auxin. The plants requiring only auxin must have a high endogenous cytokinin level in their meristems. Among auxins, the use of 2,4-D should be avoided which promotes only callusing. NAA and IAA are widely used auxins and NAA being preferred due to better stability. The role of GA_3 is also emphasized by few authors which is suggested to promote better growth and differentiation and suppresses callusing from meristem explants. Both liquid and semi-solid media have been tried for meristem–tip culture but, agar medium is generally preferred.

Explant Size

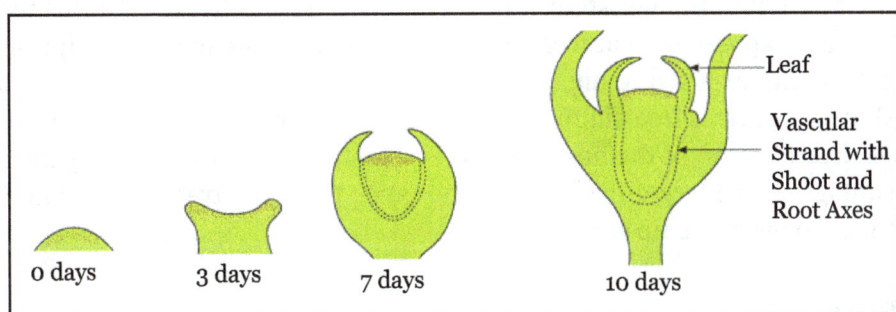

Schematic representation of development of bipolar axes by meristem culture.

The survival of the meristem tips, under the controlled condition, is determined by the size of the explant. The larger the explant, the greater are the chances of plant regeneration. However, the survival of the explants can not be treated independent of the efficiency with which virus elimination is achieved that is inversely related to the size of the explant. Thus, explants should be small enough to eradicate viruses and large enough to be able to develop into a complete plant. Besides the size of the explant, the presence of leaf primordia influences the ability of the meristems to form plants. In some plants it is essential to excise shoot meristems with two to three leaf primordia. Smith and

Murashige have suggested that leaf primordia supply auxin and cytokinin to the meristem necessary for its growth and differentiation. In a culture medium containing essential growth regulators, the excised meristems domes develop bipolar axes very quickly during reorganization. Once the root-shoot axis is established further development follows the same pattern as that of seedlings.

Storage Conditions

Generally, light incubation of meristem tip culture is found better than dark incubation. The light intensity could range from 100 lx to 4000 lx which should increase in succession as the differentiation of meristem explant progresses. There are no clear information on the effect of temperature on regeneration of plants from excised meristem tips. The cultures are normally stored under room temperature (25 ± 2 °C) conditions.

Physiological Conditions of the Explant

Meristem-tips should be collected from actively growing buds. In few cases, the tips taken from terminal buds proved better than those taken from axillary buds. Seeing the higher number of axillary buds present per shoot, in majority of the reports axillary buds were utilized as explants to increase the overall production of virus-free plants. The time excision of buds is also critical, specially for the trees with periodic growth. For example, in temperate trees the growth of the plant is limited to only a very short period in the spring and afterwards dormancy starts. In such cases, the meristem-tip cultures can be raised during spring only for increased success rate.

Thermotherapy

Often, apical meristems are not always free of virus and it can't be considered as a universal occurrence. There are certain viruses like, Tobacco Mosaic Virus (TMV), Potato Virus X (PVX) and Cucumber Mosaic Virus (CMV), which invade the meristematic region of the growing tips and interrupts the growth of the meristematic tissue. In such cases also it has been possible to obtain virus-free plants by combining meristem-tip culture with thermotherapy. In this technique, first the mother plants are exposed to heat treatment before excising the meristem-tips or, alternatively, shoot-tip cultures are exposed to high temperature regimes (35 °C-40 °C) for certain duration (6h to 6 weeks) to obtain virus free plants. In the later case, continuous exposure to very high temperature causes deterioration of the host tissues. The first procedure of treating the mother plant has added advantage where larger explants can be taken from the treated stock and thus, favors relatively higher chances of the tip survival.

Chemotherapy

Chemotherapy is the treatment of an ailment by chemicals especially by killing micro-organisms. It will not eradicate the virus completely. However, a large number of antibiotics, growth regulators, amino acids, purines and pyrimidines can be tested for inactivation of viruses. A nucleotide analogue ribavirin has been found to be the most efficient viracide for plant viruses. This broad spectrum antiviral agent, effective against both plant and animal, was reported to eliminate PVY, CMV and TMV from tobacco explant cultures, Chlorotic Leaf Spot Virus (CLSV) in apple cultures when incorporated into the medium. Vidarabine (adenine arabinoside) and antiserum are also known to reduce the titre of viruses. The effectivity of the compound may vary with the virus and the host genotype.

Virus Elimination through Callus Culture

It is a general observation that not all the cells in a calli uniformly carry the pathogen when raised from infected tissues. The two possible reasons for the escape of some cells of a systematically infected callus from virus infection are: (a) virus replication is unable to keep pace with cell proliferation, and (b) some cells acquire resistance to virus infection through mutagenesis. Therefore, it is possible to raise virus-free plants from infected shoot-tip calli. However, genetic instability of cultured cells and lack of plant regeneration in callus cultures of some plants poses the limitations of using calli for virus elimination.

Virus Indexing

Even after subjecting the meristem-tips to various treatments favoring virus eradication, only a proportion of the cultures yield virus free plants. Therefore, it is required to test all plants, regenerated through meristem-tip or callus cultures, for specific viruses before being used as mother plant to produce virus-free stock. The individual plants consistently showing negative results for virus titre can be marked as 'virus tested' for specific virus/es and can be released for commercial purposes. The following tests can be performed for virus testing:

1. The simplest test for the presence or absence of viruses in plant tissues is to examine the leaves and stem for the visible symptoms characteristic of the virus.

2. Another test is the sap transmission test or 'bioassay test' or 'infectivity test'. It is a very sensitive test and can be performed at a commercial scale. To perform this, ground the test leaves in equal volume (w/v) of 0.5M phosphate buffer using a mortar and pestle. Leaves of the indicator plant (a plant very susceptible to specific viruses), dusted with 600-grade carborundum, are swabbed with the leaf sap from the test plant. After 5 min the incubated leaves are gently washed with water to remove the residual inoculum. The inoculated indicator plants are maintained in a glasshouse, separate from other plants. It may take several days to several weeks, depending on the nature of virus and the virus titre, for the symptoms to appear on the indicator plants. It is used to detect some viruses and viroids but is a slow process requiring several days to months.

3. The third method, enzyme-linked immunosorbant assay (ELISA), is more rapid serological test which allows quick detection of important viruses. It relies on the use of antibodies prepared against the viral coat protein, requires only a small amount of antiserum and can be performed with simple equipment. However, it is not applicable to viroids and viruses which have lost their coat proteins.

Plant Regeneration

The phenomena of regeneration in plants have been known much longer than those in animals. For cuttings of twigs and the separated leaves of many plants to become independent individuals, or to give rise to them, is an every-day occurrence. The anatomical changes which take place are, in general, well known. On the other hand, relatively few investigations have been made as to the factors which set up regeneration and determine the kind of organs and the manner of their formation in regeneration. It is evident without further discussion that a knowledge of these facts would be of the greatest importance for every theory of organic development and heredity, and in brief for all investigations which might be classed under "causal morphology." It is necessary to group the facts from some general poinlt of view before a theory can be formulated.

The phenomena of regeneration imply a development of dormant or latent rudiments. These rudiments (Anlagen) are present as vegetative points (embryonic tissue) and are set into activity by injuries, or they are outwardly invisible, there being simply a disposition or tendency toward the formation of new structures, as in adventitious buds, or adventitious roots. The two cases are not sharply distinguishable from each other, since in both the unfolding of a rudiment, or the awakening of a pre-disposition, is conditioned by the reciprocal connections of organs with one another, which are designated as "correlations."

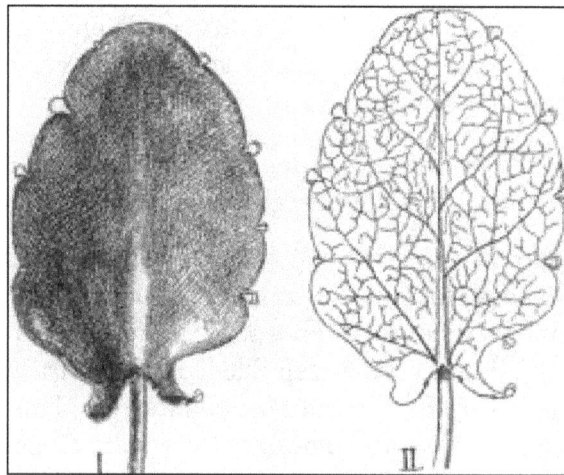

Bryophyflzm crenatum. Detached leaf, which has developed sprouts upon its edge, each with two small leaves. (In 11 made translucent so that the venation is visible).

Some examples will make this point clearer. It is well known that every tree has hundreds of dormant buds which ordinarily do not awake, but which may be set into activity by cutting back, or by the destruction of the leaves during the vegetative season. My investigations were most closely concerned with the development of the buds which are normally present on the leaves of some Crassulaceae, as for example Bryopityllum crenatum, and which are laid down even in the embryonic condition of the leaves. Their presence implies that the leaves here serve the function of reproduction, since every leaf that is cut off, if laid in damp earth, produces numerous young plants in the notches of its edges.

The Range of Regenerative Capability in Plants

Virtually no group of organisms lacks the ability to regenerate something. This process, however, is developed to a remarkable degree in lower organisms, such as protists and plants, and even in many invertebrate animals such as earthworms and starfishes. Regeneration is much more restricted in higher organisms such as mammals, in which it is probably incompatible with the evolution of other body features of greater survival value to these complex animals.

Protists and Plants

Algae

One of the most outstanding feats of regeneration occurs in the single-celled green alga Acetabularia. This plant-like protist of shallow tropical water consists of a group of short rootlike appendages; a long thin "stem," up to several centimetres in length; and an umbrella-like cap at the top. The entire organism is one cell, with its single nucleus situated at the base in one of the "roots." If the cap is cut off, a new one regenerates from the healed over stump of the amputated stem. The nucleus is necessary for this kind of regeneration, presumably because it provides the information needed to direct the development of the new cap. Once this information has been produced by the nucleus, however, the nucleus can be removed and regeneration continues unabated.

If the nucleus from one species of Acetabularia is added to a cell-body of another species, and the cap of the recipient cell is amputated, the new cap that regenerates will be a hybrid because each nucleus exerts its own morphogenetic influences. On the other hand, if the nucleus from one species is substituted for that in another, regeneration reflects the properties of the new nucleus.

Protozoans

Most single-celled, animal-like protists regenerate very well. If part of the cell fluid, or cytoplasm, is removed from Amoeba, it is readily replaced. A similar process occurs in other protozoans, such as flagellates and ciliates. In each case, however, regeneration occurs only from that fragment of the cell containing the nucleus. Amputated parts that lack a nucleus cannot survive. In some ciliates, such as Blepharisma or Stentor, the nucleus may be elongated or shaped like a string of beads. If either of these organisms is cut in two so that each fragment retains part of the elongated nucleus, each half proceeds to grow back what it lacks, giving rise to a complete organism in less than six hours. The way in which such a bisected protozoan regenerates is almost identical with the way it reproduces by ordinary division. Even a very tiny fragment of the whole organism can regenerate itself, provided it contains some nuclear material to determine what is supposed to be regenerated.

Green Plants

The mechanisms by which vascular plants grow have much in common with regeneration. Their roots and shoots elongate by virtue of the cells in their meristems, the conical growth buds at the tip of each branch. These meristems are capable of indefinite growth, especially in perennial plants. If they are amputated they are not replaced, but other meristems along the stem, normally held in abeyance, begin to sprout into new branches that more than compensate for the loss of the original one. Such a process is called restitution.

Plants are also capable of producing callus tissue wherever they may be injured. This callus is proliferated from cambial cells, which lie beneath the surface of branches and are responsible for their increase in width. When a callus forms, some of its cells may organize into growing points, some of which in turn give rise to roots while others produce stems and leaves.

Production of Haploid Plants

Haploid plants are characterized by possessing only a single set of chromosomes (gametophytic number of chromosomes i.e. n) in the sporophyte. This is in contrast to diploids which contain two sets (2n) of chromosomes. Haploid plants are of great significance for the production of homozygous lines (homozygous plants) and for the improvement of plants in plant breeding programmes.

Grouping of Haploids

Haploids may be divided into two broad categories:

Monoploids (Monohapioids)

These are the haploids that possess half the number of chromosomes from a diploid species e.g. maize, barley.

Polyhaploids

The haploids possessing half the number of chromosomes from a polyploid species are regarded as polyhaploids e.g. wheat, potato. It may be noted that when the term haploid is generally used it applies to any plant originating from a sporophyte (2n) and containing half the number (n) of chromosomes.

In Vivo and In Vitro Approaches

The importance of haploids in the field of plant breeding and genetics was realised long ago. Their practical application, however, has been restricted due to very a low frequency (< 0.001%) of their formation in nature.

The process of apomixis or parthenogenesis (development of embryo from an unfertilized egg) is responsible for the spontaneous natural production of haploids. Many attempts were made, both by in vivo and in vitro methods to develop haploids. The success was much higher by in vitro techniques.

In Vivo Techniques for Haploid Production

There are several methods to induce haploid production in vivo.

Some of them are listed below:

Androgenesis

Development of an egg cell containing male nucleus to a haploid is referred to as androgenesis. For a successful in vivo androgenesis, the egg nucleus has to be inactivated or eliminated before fertilization.

Gynogenesis

An unfertilized egg can be manipulated (by delayed pollination) to develop into a haploid plant.

Distant Hybridization

Hybrids can be produced by elimination of one of the parental genomes as a result of distant (interspecific or inter-generic crosses) hybridization.

Irradiation Effects

Ultra violet rays or X-rays may be used to induce chromosomal breakage and their subsequent elimination to produce haploids.

Chemical Treatment

Certain chemicals (e.g., chloramphenicol, colchicine, nitrous oxide, maleic hydrazide) can induce chromosomal elimination in somatic cells which may result in haploids.

In Vitro Techniques for Haploid Production

In the plant biotechnology programmes, haploid production is achieved by two methods.

Androgenesis

Haploid production occurs through anther or pollen culture, and they are referred to as androgenic haploids.

Gynogenesis

Ovary or ovule culture that results in the production of haploids, known as gynogenic haploids.

Androgenesis

In androgenesis, the male gametophyte (microspore or immature pollen) produces haploid plant. The basic principle is to stop the development of pollen cell into a gamete (sex cell) and force it to develop into a haploid plant. There are two approaches in androgenesis— anther culture and pollen (microspore) culture. Young plants, grown under optimal conditions of light, temperature and humidity, are suitable for androgenesis.

Anther Culture

The selected flower buds of young plants are surface-sterilized and anthers removed along with their filaments. The anthers are excised under aseptic conditions, and crushed in 1% acetocarmine to test the stage of pollen development.

If they are at the correct stage, each anther is gently separated (from the filament) and the intact anthers are inoculated on a nutrient medium. Injured anthers should not be used in cultures as they result in callusing of anther wall tissue.

The anther cultures are maintained in alternating periods of light (12-18 hr.) and darkness (6-12 hrs.) at 28 °C. As the anthers proliferate, they produce callus which later forms an embryo and then a haploid plant.

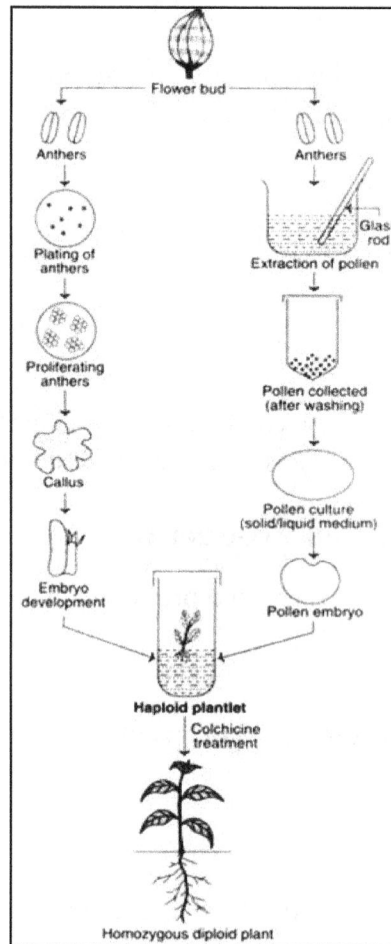

Diagrammatic representation of anther and pollen cultures
for the production of haploid and diploid plants.

Pollen (Microspore) Culture

Haploid plants can be produced from immature pollen or microspores (male gametophytic cells). The pollen can be extracted by pressing and squeezing the anthers with a glass rod against the sides of a beaker. The pollen suspension is filtered to remove anther tissue debris.

Viable and large pollen (smaller pollen do not regenerate) are concentrated by filtration, washed and collected. These pollen are cultured on a solid or liquid medium. The callus/embryo formed is transferred to a suitable medium to finally produce a haploid plant, and then a diploid plant (on colchicine treatment).

Comparison between Anther and Pollen Cultures

Anther culture is easy, quick and practicable. Anther walls act as conditioning factors and promote culture growth. Thus, anther cultures are reasonably efficient for haploid production. The major limitation is that the plants not only originate from pollen but also from other parts of anther. This results in the population of plants at different ploidy levels (diploids, aneuploids). The disadvantages associated with anther culture can be overcome by pollen culture.

Many workers prefer pollen culture, even though the degree of success is low, as it offers the following advantages:

1. Undesirable effects of anther wall and associated tissues can be avoided.

2. Androgenesis, starting from a single cell, can be better regulated.

3. Isolated microspores (pollen) are ideal for various genetic manipulations (transformation, mutagenesis).

4. The yield of haploid plants is relatively higher.

Development of Androgenic Haploids

The process of in vitro androgenesis for the ultimate production of haploid plants is depicted in figure.

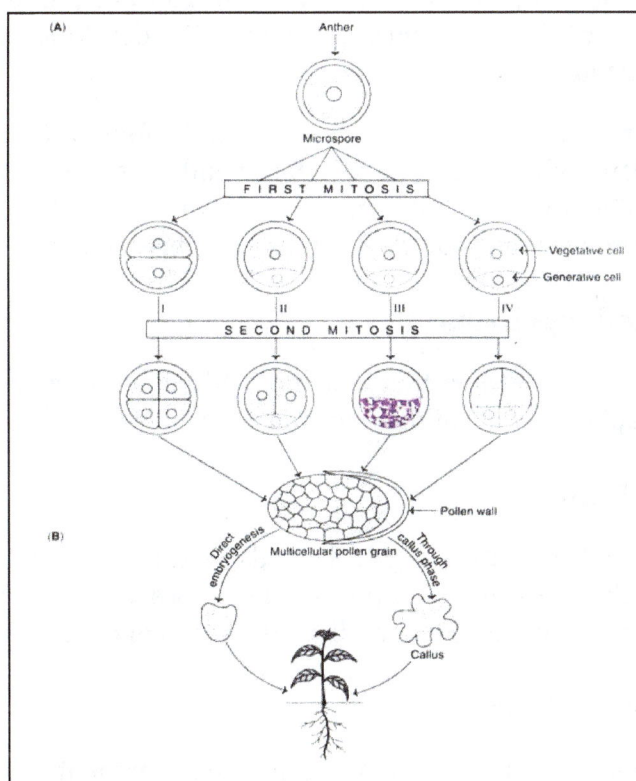

Diagrammatic representation of microscope divisions leading to the formation of a multicellular pollen grain (A), followed by the formation of haploid scorophyte (B) (Note : I, II, III and IV indicate respective pathways).

The cultured microspores mainly follow four distinct pathways during the initial stages of in vitro androgenesis.

Pathway I

The uninucleate microspore undergoes equal division to form two daughter cells of equal size e.g. Datura innoxia.

Pathway II

In certain plants, the microspore divides unequally to give bigger vegetative cell and a smaller generative cell. It is the vegetative cell that undergoes further divisions to form callus or embryo. The generative cell, on the other hand, degenerates after one or two divisions—e.g., Nicotiana tabacum, Capsicum annuum.

Pathway III

In this case, the microspore undergoes unequal division. The embryos are formed from the generative cell while the vegetative cell does not divide at all or undergoes limited number of divisions e.g. HyoScyamus niger.

Pathway IV

The microspore divides unequally as in pathways I and II. However, in this case, both vegetative and generative cells can further divide and contribute to the development of haploid plant e.g. Datura metel, Atropa belladonna.

At the initial stages, the microspore may follow any one of the four pathways described above. As the cells divide, the pollen grain becomes multicellular and burst open. This multicellular mass may form a callus which later differentiates into a plant (through callus phase). Alternately, the multicellular mass may produce the plant through direct embryogenesis.

Factors Affecting Androgenesis

A good knowledge of the various factors that influence androgenesis will help to improve the production of androgenic haploids. Some of these factors are briefly described.

Genotype of Donar Plants

The success of anther or pollen culture largely depends on the genotype of the donor plant. It is therefore important to select only highly responsive genotypes. Some workers choose a breeding approach for improvement of genotype before they are used in androgenesis.

Stage of Microspore or Pollen

The selection of anthers at an ideal stage of microspore development is very critical for haploid production. In general, microspores ranging from tetrad to bi-nucleate stages are more responsive. Anthers at a very young stage (with microspore mother cells or tetrads) and late stage (with bi-nucleate microspores) are usually not suitable for androgenesis. However, for maximum production of androgenic haploids, the suitable stage of microspore development is dependent on the plant species, and has to be carefully selected.

Physiological Status of a Donar Plant

The plants grown under best natural environmental conditions (light, temperature, nutrition, CO_2 etc.) with good anthers and healthy microspores are most suitable as donor plants. Flowers obtained from young plants, at the beginning of the flowering season are highly responsive. The use of pesticides should be avoided at least 3-4 weeks preceding sampling.

Pretreatment of Anthers

The basic principle of native androgenesis is to stop the conversion of pollen cell into a gamete, and force its development into a plant. This is in fact an abnormal pathway induced to achieve in vitro androgenesis. Appropriate treatment of anthers is required for good success of haploid production. Treatment methods are variable and largely depend on the donor plant species.

Chemical Treatment

Certain chemicals are known to induce parthenogenesis e.g. 2-chloroethylphosphonic acid (ethrel). When plants are treated with ethreal, multinucleated pollens are produced. These pollens when cultured may form embryos.

Temperature Influence

In general, when the buds are treated with cold temperatures (3-6 °C) for about 3 days, induction occurs to yield pollen embryos in some plants e.g. Datura, Nicotiana. Further, induction of androgenesis is better if anthers are stored at low temperature, prior to culture e.g. maize, rye. There are also reports that pretreatment of anthers of certain plants at higher temperatures (35 °C) stimulates androgenesis e.g. some species of Brassica and Capsicum.

Effect of Light

In general, the production of haploids is better in light. There are however, certain plants which can grow well in both light and dark. Isolated pollen (not the anther) appears to be sensitive to light. Thus, low intensity of light promotes development of embryos in pollen cultures e.g. tobacco.

Effect of Culture Medium

The success of another culture and androgenesis is also dependent on the composition of the medium. There is, however, no single medium suitable for anther cultures of all plant species. The commonly used media for anther cultures are MS, White's, Nitsch and Nitsch, N6 and B5. These media in fact are the same as used in plant cell and tissue cultures. In recent years, some workers have developed specially designed media for anther cultures of cereals.

Sucrose, nitrate, ammonium salts, amino acids and minerals are essential for androgenesis. In some species, growth regulators — auxin and/or cytokinin are required for optimal growth. In certain plant species, addition of glutathione and ascorbic acid promotes androgenesis. When the anther culture medium is supplemented with activated charcoal, enhanced androgenesis is observed. It is believed that the activated charcoal removes the inhibitors from the medium and facilitates haploid formation.

Gynogenesis

Haploid plants can be developed from ovary or ovule cultures. It is possible to trigger female gametophytes (megaspores) of angiosperms to develop into a sporophyte. The plants so produced are referred to as gynogenic haploids.

Gynogenic haploids were first developed by San Noem from the ovary cultures of Hordeum vulgare. This technique was later applied for raising haploid plants of rice, wheat, maize, sunflower, sugar beet and tobacco.

In vitro culture of un-pollinated ovaries (or ovules) is usually employed when the anther cultures give .unsatisfactory results for the production of haploid plants. The procedure for gynogenic haploid production is briefly described.

The flower buds are excised 24-48 hr. prior to anthesis from un-pollinated ovaries. After removal of calyx, corolla and stamens, the ovaries are subjected to surface sterilization. The ovary, with a cut end at the distal part of pedicel, is inserted in the solid culture medium.

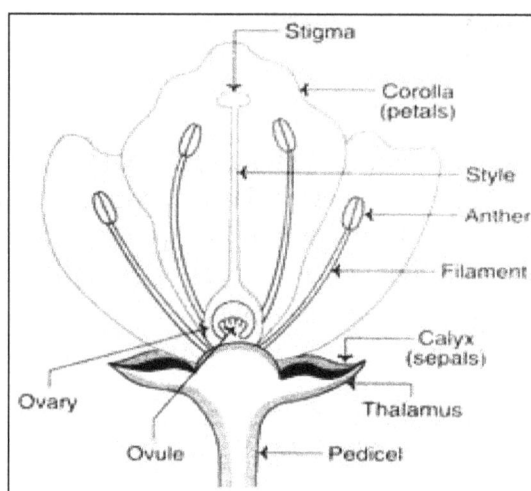
A diagrammatic representation of important parts in flower.

Whenever a liquid medium is used, the ovaries are placed on a filter paper or allowed to float over the medium with pedicel inserted through filter paper. The commonly used media are MS, White's, N6 and Nitsch, supplemented growth factors. Production of gynogenic haploids is particularly useful in plants with male sterile genotype. For such plant species, this technique is superior to another culture technique.

Limitations of Gynogenesis

In practice, production of haploid plants by ovary/ovule cultures is not used as frequently as anther/pollen cultures in crop improvement programmes.

The major limitations of gynogenesis are listed:

1. The dissection of unfertilized ovaries and ovules is rather difficult.

2. The presence of only one ovary per flower is another disadvantage. In contrast, there are a large number of microspores in one another.

However, the future of gynogenesis may be more promising with improved and refined methods.

Identification of Haploids

Two approaches based on morphology and genetics are commonly used to detect or identify haploids.

Morphological Approach

The vegetative and floral parts and the cell sizes of haploid plants are relatively reduced when compared to diploid plants. By this way haploids can be detected in a population of diploids. Morphological approach, however, is not as effective as genetic approach.

Genetic Approach

Genetic markers are widely used for the specific identification of haploids. Several markers are in use.

1. 'a$_1$' marker for brown coloured aleurone.

2. 'A' marker for purple colour.

3. 'Lg' marker for ligule less character.

The above markers have been used for the development of haploids of maize. It may be noted that for the detection of androgenic haploids, the dominant gene marker should be present in the female plant.

Diploidizatioim of Haploid Plants (Production of Homozygous Plants)

Haploid plants are obtained either by androgenesis or gynogenesis. These plants may grow up to a flowering stage, but viable gametes cannot be formed due to lack of one set of homologous chromosomes. Consequently, there is no seed formation.

Haploids can be diploidized (by duplication of chromosomes) to produce homozygous plants. There are mainly two approaches for diploidization— colchicine treatment and endomitosis.

Colchicine Treatment

Colchicine is very widely used for diploidization of homologous chromosomes. It acts as an inhibitor of spindle formation during mitosis and induces chromosome duplication. There are many ways of colchicine treatment to achieve diploidization for production of homozygous plants.

1. When the plants are mature, colchicine in the form of a paste is applied to the axils of leaves. Now, the main axis is decapitated. This stimulates the axillary buds to grow into diploid and fertile branches.

2. The young plantlets are directly treated with colchicine solution, washed thoroughly and replanted. This results in homozygous plants.

3. The axillary buds can be repeatedly treated with colchicine cotton wool for about 2-3 weeks.

Endomitosis

Endomitosis is the phenomenon of doubling the number of chromosomes without division of the nucleus. The haploid cells, in general, are unstable in culture with a tendency to undergo endomitosis. This property of haploid cells is exploited for diploidization to produce homozygous plants.

The procedure involves growing a small segment of haploid plant stem in a suitable medium supplemented with growth regulators (auxin and cytokinin). This induces callus formation followed by differentiation. During the growth of callus, chromosomal doubling occurs by endomitosis. This results in the production of diploid homozygous cells and ultimately plants.

Applications of Haploid Plants

Development of Homozygous Lines

It is now possible to develop homozygous lines within a span of few months or a year by employing anther/pollen culture. This is in contrast to the conventional plant breeding programme that might take several years (6-10 yrs). In this way, production of haploids is highly useful for research related to plant genetics and breeding.

Generation of Exclusive Male Plants

By the process of androgenesis to produce haploids, followed by chromosome doubling, it is possible to develop exclusive male plants. The male plants are particularly useful when their productivity and applications are much more than female plants.

Induction of Mutations

In general, majority of induced mutations are recessive and therefore are not expressed in diploid cells (due to the presence of dominant allele). Haploids provide a convenient system for the induction of mutations and selection of mutants with desired traits. In fact, the haploid cells can be cultured and handled in a fashion similar to microorganisms.

Mutants from several plant species that are resistant to antibiotics, toxins, herbicides etc. have been developed. When the haploid cells of tobacco plant (Nicotiana tabacum) were exposed to methionine sulfoximine (a mutagen), mutants which showed lower level of infection to Pseudomonas tabaci were produced.

Production of Disease Resistance Plants

Disease resistance genes can be introduced while producing haploids. The so developed haploids are screened for the desired resistance, and then diploidized.

Some examples of disease resistance plants are listed:

1. Hwansambye, a rice variety resistant to leaf blast, bacterial leaf blight and rice stripe tenui virus.

2. Barley accession Q21681 resistant to stem rust, leaf rust and powdery mildew.

More examples of disease resistance crops are given in table below.

Table: A selected list of Improved varieties of crops developed by using anther culture.

Crop	Varieties	Improvements made
Wheat (Triticum aestivum)	Lunghua 1, Zing Hua 1, Zing Hua 2 Huapei 1, Florin, Ambitus, Jingdan 2288 .	High yield, rust resistance, cold resistance, large spikes, more tillers.
Rice (Oryza sativa)	Tangfong 1, Xin Xiu, Zhog Hua 8 Zhong Hua 9, Hua yu 1, Hua Yu 2, Huapei Shanyou 63, Zhe keng 66, Ta Be 78, Nonhua 5, Hirohikari, Hirohonami.	High, yield, good quantity, disease resistance.
Tobacco (Nicotiana tavacum)	Tanyu 1, Tanyu 2, Tanyu 3, F 211 Hai Hua 19, Hai Hua 30.	Mild smoking, disease resistance.
Brassica napus	Jai Kisam	Low erucic acid

Production of Insect Resistance Plants

Some varieties of rice resistant to insects have been developed e.g. Hwacheongbyeo resistant to brown plant hopper. Other varieties of rice that are resistant to pests have also been produced.

Production of Salt Tolerance Plants

The plant species with salt tolerance are needed for their cultivation in some areas. Anther cultures have resulted in some varieties of rice and wheat with good salt tolerance.

Cytogenetic Research

Haploids are useful in several areas of cytogenetic research. These include:

1. Production of aneuploids,

2. Determination of the nature of ploidy,

3. Determination of basic chromosome number,

4. Evaluation of origin of chromosomes.

Induction of Genetic Variability

Besides the development of haploid mutants, it is also possible to produce plants with various ploidy levels through androgenesis.

Doubled Haploids in Genome Mapping

Genome mapping, a recent development in molecular biology, can be more conveniently achieved by using doubled haploid plant species.

Evolutionary Studies

A comparison of di-haploids (doubled haploids) with diploid wild plant species will be useful to trace the evolutionary origin of various plants. The close evolutionary relationship between tomato and potato has been evaluated by this approach.

Crops Developed through Anther Cultures

Anther culture has made it possible to develop many new varieties of plants; a selected list is given in table. In general, the new crops are high-yielding varieties with disease resistance. Some of them are resistance to cold, salt etc. Many plant breeders these days use anther cultures in their regular breeding programmes.

Triploid Production by Endosperm Culture

Endosperm is a unique tissue in its origin, development and ploidy level. It is a product of double fertilization but unlike the embryo it is triploid and develops into a formless tissue. It is, therefore, an interesting tissue for morphogenesis. Any abnormality in the development of endosperm may cause the abortion of embryo resulting in sterile seeds. The endosperm may be totally consumed by developing embryo leading to the formation of exalbuminous (non-endospermous) seeds or when it persists, the seed is called albuminous (endospermous). In albuminous seeds, it is used as a food source which may contain proteins, starch or fats and the embryo can utilize this food during seed germination.

Cellular totipotency of endosperm cells was first demonstrated by Johri and Bhojwani in 1965. To date, differentiation of shoots/embryos/plantlets from endosperm tissue has been reported for more than 64 species belonging to 24 families. In many of these reports the regenerants were shown to be triploid. A key factor in the induction of cell divisions in mature endosperm cultures is the association of embryo. The embryo factor is required only to trigger cell divisions; further growth occurs independent of the embryo. Triploid plants are usually seed-sterile. However, there are many examples where seedlessness caused by triploidy is of no serious concern or, at times, even advantageous. Some of the crops where triploids are already in commercial use include several varieties of apple, banana, mulberry, sugar beet and watermelon. Natural triploids of tomato produced larger and tastier fruits than their diploid counterparts.

Traditionally, triploids are produced by crossing induced superior tetraploids and diploids. This approach is not only tedious and lengthy (especially for tree species) but in many cases it may not be possible due to high sterility of autotetraploids. The first step in the process is to produce tetraploids by colchicine treatment of germinating seeds, seedlings or vegetative buds. In most of these cases the rate of induction of tetraploids had been low (7-22%). Moreover, the treatment is lengthy and laborious. Once tetraploids have been produced, their cross with the diploid parent may not be successful in majority of the cases. In successful crosses the seed-set, seed germination and survival rate of the seedlings is low. Moreover, all sexually produced triploids do not behave uniformly, which may be due to segregation both at tetraploid level and subsequent population of crosses with putative diploid. In contrast, in vitro regeneration of plants from endosperm, a naturally occurring triploid tissue, offers a direct, single step approach to triploid production. The selected triploids, expected to be sexually sterile, can be bulked up by micropropagation.

Factors Controlling Callus Proliferation and Plant Regeneration

Endosperm at Culture

Usually culture of endosperm needs the selection of seeds at proper stage of development. This is usually calculated as days after pollination (DAP) and it varies from plant to plant as 9-10 days after pollination (DAP) in Lolium perenne, 8-11 DAP in Zea mays, 8 DAP in Triticum aestivum and Hordeum vulgare and 4-7 DAP in Oryza sativa. Usually free nuclear endosperm did not produce any callus and the intensity of response depends on the level of organization of endosperm cells.

Plant Growth Regulators and other Supplements

Selection of a suitable basal medium and the addition of proper growth regulators and other adjuvants are the decisive factors that determine the success of triploid plant development. The culture of immature endosperm requires yeast extract (YE), casein hydrolysate (CH), coconut milk (CM), corn extract (CE), potato extract (PE), grape juice (GJ), cow's milk (CWM) or tomato juice (TJ) despite a suitable medium and growth regulators. Murashige and Skoog basal medium was mostly used to initiate and improve the response in in vitro endosperm cultures.

Most of the immature endosperm needs the presence of one or more growth regulators for plant regeneration except in few , where MS basal medium is sufficient for endosperm embryogenesis. In majority of reports, an auxin, preferably 2,4-D is necessary for callus induction from immature endosperms. In case of mature endosperm, the optimum callus growth was observed either on a cytokinin or a cytokinin in combination with an auxin and for autotrophic taxa, cytokinin, auxin, CH or YE is necessary. In most of the cases the time required to initiate proliferation varies from 10 days to 20 days, but pre-soaking of endosperms with GA_3 have reduced the time period from 10 days to 7 days.

Physical Factors

This includes effect of temperature, light and pH on endosperm proliferation. In some cases, the endosperms were cultured along with the embryo and kept in the diffuse light with 16 h photoperiod. Light conditions facilitate the early germination of embryo and the embryos can be removed easily due to their characteristic green colour. In coffee, the endosperm callus grows better under 12 h light/dark conditions. In Lolium the light doesn't have any significant role on endosperm proliferation.

Not much research has been carried out till date with regard to the effect of temperature and pH on endosperm proliferation. In available literature the optimum temperature for endosperm growth was reported to be 25 °C. The pH varies from 4.0 for Asimina to 5.0 for Ricinus , 5.6 for Jatropa and Putranjiva and 6.1 for Zea mays. In general, 5.5 to 5.8 pH seem to support the best growth of endosperm tissues in cultures.

The Embryo Factor

There is an absolute necessity of the so called "embryo factor" for the proliferation of endosperm. Some factors contributed by the germinating embryo is required for the stimulation of mature and dried endosperms of few plant species. In general, it has been found that mature endosperm requires the initial association of embryo to form callus but immature endosperm proliferates independent of the embryo. However, in neem the association of the embryo proved essential to induce callusing of immature endosperm; the best explant was immature seeds. Similar observation for mulberry was reported by Thomas et al.. However, the embryo factor can be overcome by the use of GA_3 as was observed in Croton bonplandianum and Putranjiva roxburghii. It is reported that during germination, the embryo releases certain gibberellin like substances, which may promote the endosperm proliferation. However the mature endosperms of Achras sapota, Santalum album, Emblica officinalis and Juglans regia could proliferate without the association of embryo or pre-soaking of them in GA_3.

Shoot Regeneration

Organogenesis from endosperm tissue was first reported in Exocarpus cupressiformis (a member of the family Santalaceae) by Johri and Bhojwani. The pathway of plant regeneration includes shoot-bud differentiation or embryogenesis directly from the explants or indirectly from proliferating callus. In almost all the parasitic angiosperms, the endosperm shows a tendency of direct differentiation of organs without prior callusing, whereas in the autotrophic taxa the endosperm usually forms callus tissue followed by the differentiation of shoot buds, roots or embryos. Direct shoot regeneration from the cultured endosperm was observed in a number of semiparasitic angiosperms including Exocarpus, Taxillus, Leptomeria, Scurrula and Dendrophthoe.

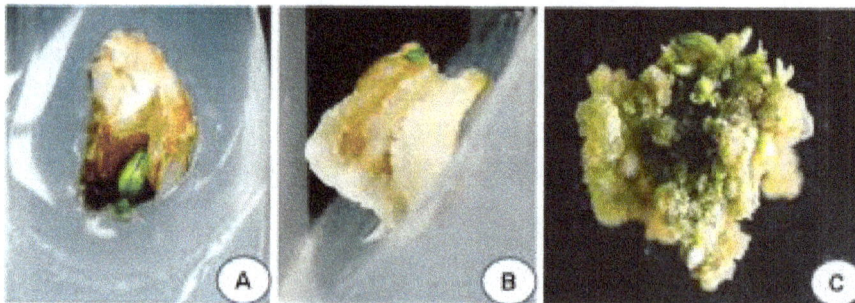

Shoot regeneration from endosperm callus of Azadirachta indica : A. An immature seed in culture has split open after 2 weeks and releasing the green embryo and callused endosperm , B. White fluffy endosperm callus can be seen from the fully opened seed after three weeks. Embryo is lying at one end of the explant, C. 6 -week-old subculture of endosperm callus showing the differentiation of distinct shoots and nodules as well.

In Exocarpus, an auxin (IAA) along with cytokinin (Kinetin) was required for direct shoot regeneration. Addition of zeatin in WM gave rise to green shoots from the intact seed (i.e. endosperm with embryo) culture of Scurrula pulverulenta which on subculture gave rise to characteristic haustoria. In Taxillus vestitus, shoot bud formation occurred on WM supplemented with IAA, Kinetin and CH, after seven weeks. Replacement of IAA with IBA could induce shoot regeneration in 55% cultures and haustoria in 60% cultures. Here, the embryo had an adverse effect on bud differentiation from endosperm. Injury to the endosperm was found to be beneficial for shoot induction in some cases; shoot buds first develop along the injured region. The position of the explant on medium plays a significant role in regeneration of shoot in Taxillus spp. When half split T. vestitus endosperm without embryo was placed on medium with its cut surface in contact with the medium containing Kinetin, 100 % of the cultures produced shoots. In Leptomeria acida, IBA proved more efficient than IAA in terms of rapid callus proliferation. However, on IAA supplemented medium the callus gave rise to shoots in 100% cultures.

The callus proliferation from endosperm and the subsequent shoot organogenesis was also reported in Jatropa panduraefolia, Putranjiva roxburghii, Codiaeum variegatum, Malus pumila, Oryza sativa, Annona squamosa, Actinidia chinensis, Mallotus philippensis, Actinidia deliciosa, Morus alba, Azadirachta indica and Actinidia deliciosa. In Actinidia species callus initiation occurred on MS medium supplemented with 2,4-D and Kinetin. Transfer of these calli to MS medium containing IAA and 2ip resulted in shoot and root organogenesis. In apple, endosperm proliferated into callus on MS medium supplemented with Kinetin + 2,4-D/BA + NAA and subsequent regeneration occurred on MS medium fortified with BA + CH. In Annona squamosa the callusing of endosperm occurred on WM supplemented with two cytokinins (Kinetin and BA), an auxin (NAA) and Gibberellic acid (GA_3). But organogenesis in the callus occurred on Nitsch's medium supplemented with BA and NAA.

In Mallotus philippensis, a continuously growing callus was obtained on MS medium supplemented with 2, 4-D + Kinetn. These calli when subcultured on MS + BA + CH gave rise to various morphologically distinct cell lines, of these, only the green compact cell line was responsive for organogenic differentiation. Shoot regeneration occurred in this callus when subcultured on MS medium fortified with BA + NAA.

In Rice, there was a striking difference in the growth response of immature and mature endosperm. Immature endosperm underwent two modes of differentition i.e. direct regeneration from the explant or indirectly via intervening callus phase. In mature endosperm, shoot organogenesis was always preceded by callusing. Callus from mature endosperm was initiated and maintained on MS + 2,4-D; shoot differentiation from callus occurred on MS + IAA + Kinetin. The proliferation of immature endosperm and occasional shoot formation occurred on YE supplemented medium; addition of IAA and Kinetin improved the response further.

Immature endosperms of neem (Azadirachta indica) showed best callusing on MS + NAA + BA + CH. When the callus was transferred to a medium containing BA or Kinetin, shoot buds differentiated from all over the callus. Maximum regeneration in terms of number of cultures showing shoot-buds and number of buds per callus cultures occurred in the presence of BA. Thomas et al. observed maximum callusing of Mulberry endosperm on MS + BA + NAA + CH or YE. Shoot buds were emerged when the callus was transferred on a medium containing a cytokinin or a cytokinin and a - naphthaleneacetic acid (NAA). The percent response was highest on BA and NAA containing medium. However, the number of shoots per explants was maximum when TDZ alone was used.

Histology

Histological studies of the proliferating endosperm of Jatropa, Putranjiva and Ricinus, revealed that the embryo also enlarged and proliferated along with the endosperm but soon showed the sign of degeneration. In such cases the endosperm calli were transferred to a fresh medium to avoid any contamination from degenerated embryonal cells. The 4-week-old callus derived from endosperm cultures, proliferated into parenchymatous cells and 6-week old callus showed tracheidal cells. In Santalum, endosperm proliferation started after the formation of several meristematic layers below the epidermal region. By carefully applying plant growth regulators the nodular outgrowths can be induced on the surface of the cultured endosperm as in case of osyris wightians and Putranjiva roxburghii.

The importance of tracheidal differentiation in the callus of endosperm cultures is that it facilitates organogenic differentiation. In the families like Euphorbiaceae, Loranthaceae and Santalaceae, the endosperm tissues readily form tracheidal elements in cultures. In Emblica officinale, tracheidal cells and cambium like cells organized into vascular strands or nodules in the differentiation medium while in the callusing medium tracheidal cells remained scattered. The differentiation of vascular strand in the callus accompanied the shoot bud formation.

In Aleuritus fordii, callus proliferated from endosperm explant consisted of large, compact and vacuolated cells. Tiny group of cells became distinct from adjoining large and vacuolated cells and became meristematic. These cells remained thin walled with dense cytoplasm and a large clear nucleus. Later the meristemoids developed in to dome shaped shoot apex, which produces leaf primordia. In Mallotus phillippensis only the compact green callus underwent differentiation. Such callus showed vasculature, developed protuberances and eventually gave rise to shoots buds. Small group of cells with deep-seated distinct meristematic loci were also observed in these calli, which later gave rise to dome-shaped shoot primordia, endogenously.

In mulberry (Morus alba), histological analysis revealed that the older region of the callus comprised of highly vacuolated cells. Shoot buds differentiated from peripheral nodular structures, which comprised of compactly arranged highly cytoplasmic cells. Often a few layers of degenerating vacuolated cells were seen outside the shoot primordia. It is possible that the shoots originated from inside the nodules and emerged after rupturing the surrounding tissue. The regenerating shoots showed vascular supply continuous with the vasculature of the callus.

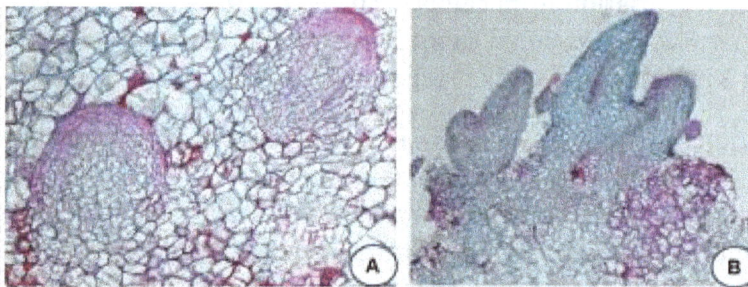

A. Section of a 2-week-old regenerating callus from endosperm of A. indica showing endogenous meristematic pockets. B. Section of a 4-week-old regenerating callus showing distinct shoot buds differentiated from peripheral vascularized nodule. One of the buds showing glands.

Both exogenous and endogenous differentiation of shoots was observed in Azadirachta indica. The serial section of two-week-old regenerating callus showed that many meristematic pockets developed from inside the callus, which developed into shoot buds after 3 weeks. Histological sections also revealed that the shoot buds emerged from the peripheral tissues of the callus as well. In Actinidia deliciosa histological analysis of the freshly isolated endosperm revealed small intercellular spaces and cells were filled with storage materials. However, the calli derived from the endosperm were larger, vacuolated and lacked storage materials. In older callus daughter nuclei attached to newly formed cell walls were often observed, suggesting disturbances of cell division. The cells differed in size and shape and contained nuclei with variable numbers of nucleoli.

Cytology

The endosperm tissue often shows a high degree of chromosomal variation and polyploidy. Mitotic irregularities, chromosome bridges and laggards are other important characteristic features of endosperm tissue. Some reports suggest that the cells of endosperm cultures showed ploidy higher than 3n as in the case of Croton, Jatropha and Lolium. Cytological observations of the endosperm callus, derived from Dendrophthoe falcata, Taxillus cuneatus and Taxillux vestitus, showed diploid (2n=18) and triploid (3n=27) chromosomes.

In addition to the cytological observations of endosperm callus, the chromosomal analysis of the regenerated plantlets were also studied in a number of systems. In Juglans regia two plants of endosperm origin were analysed for ploidy determination and both the plants showed triploid (3n=3x=48) number of chromosomes. In Citrus stability of the ploidy level and chromosome number were observed all through the regeneration process and triploid (2n=3x=27) plantlets were recovered. In Mallotus philippensis the squash preparation of root tip cells of the regenerated plants invariably showed triploid chromosome number (3n=3x=33). The triploid nature of the endosperm-derived plants was determined by Feulgen cytophotometry in Acacia nilotica.

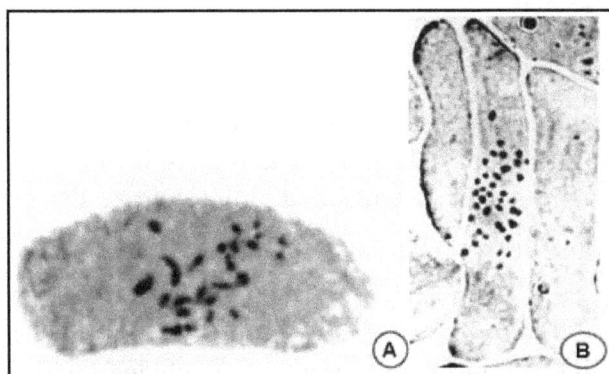

Cells from the root-tips of shoots of endosperm origin: A. showing diploid number of chromosomes (2n=2x=24), B. triploid number of chromosomes (2n=3x=36).

In Mulberry (Morus alba), 7-month-old plants of endosperm origin were utilized for ploidy determination. All the ten plants analysed cytologically showed triploid number of chromosome (2n=3x=42). The ploidy determination of 20 plants of Azadirachta indica , regenerated from endosperm calli, showed that 66% of the plants had triploid number of chromosomes (2n=3x=36) and the rest 34% were diploids (2n=2x=24). In Actinidia deliciosa three different ploidy levels viz., 3C, 6C and 9C were observed in cells of endosperm derived callus analyzed by flow cytometry. The analysis of the leaves of endosperm derived plants showed 45.7% fluorescence intensity peaks corresponding exactly to 3C whereas 42.2% exhibited peaks of fluorescence intensity representing C-values between 2C and 4C. Only 8.4% of the samples indicated 2C DNA content, and one sample showed 6C DNA content.

Germplasm Conservation

Germplasm conservation is the most successful method to conserve the genetic traits of endangered and commercially valuable species. Germplasm is a live information source for all the genes present in the respective plant, which can be conserved for long periods and regenerated whenever it is required in the future. Different plants' genetic diversity is also preserved and this in turn creates a pool of different genes that acts as resource house (gene bank/library) for new or unidentified species. This gene bank also assists the in vitro testing of germplasms for their genetic alterations (transgene) and to screen the elite germplasms among them. Genetic variation leads to the loss of genetic information of earlier generations, which makes germplasm conservation an important aspect to preserve this information regarding endangered, primitive, or even any existing species. Germplasm conservation includes:

- Breeding lines,
- Cultivated species,
- Commercial varieties,
- Special genetic stocks,
- Landraces,
- Wild species (for direct use (crop species) and indirect use (as root stocks)).

Germplasm Preservation

In-situ Conservation

On-site conservation is called as in-situ conservation, which means conservation of genetic resources in the form of natural populations by establishing biosphere reserves such as national parks and sanctuaries. Practices like horticulture and floriculture also preserve plants in a natural habitat.

Ex-situ Conservation

Off-site conservation is called as ex-situ conservation, which deals with conservation of an endangered species outside its natural habitat. In this method genetic information of cultivated and wild plant species is preserved in the form of in vitro cultures and seeds, which are stored as gene banks for long-term use. This type of conservation creates a bank of genes/DNA, seeds, and germplasms and forms a genetic information library (e.g., common garden archives) for endangered, primitive, and commercially valuable species. It also includes certain preservation (cryopreservation) and gene transformation techniques for the incorporation of disease, pest and stress tolerance traits, and environmental restoration of endangered plant species.

Genetic resources are used for a variety of reasons such as genetic improvement, conservation of biodiversity, mechanistic studies of adaptation, systematics and taxonomy, environmental monitoring, epidemiology, and forensics. One of the main strategies behind germplasm conservation is to maintain the biological integrity and provide germplasms with validated phenotypic and genetic descriptions. Gene banks are represented as in vivo and in vitro gene banks. Banks in which genetic resources are preserved by conventional methods, for example, seeds, vegetative propagules, etc., are called in vivo gene banks, whereas banks in which the genetic resources are preserved by nonconventional methods such as cell and tissue culture methods are called in vitro gene banks. Both these ensure the availability of valuable germplasms to a breeder to develop new and improved varieties.

Methods Involved in the In Vitro Conservation of Germplasm

Cryopreservation

In this technique (Greek krayos, meaning frost) the cells or tissues are preserved in a frozen state at very low temperatures using solid carbon dioxide (at -79 °C), with low temperature deep freezers (at -80 °C), using vapor nitrogen (at -150 °C) and liquid nitrogen (at -196 °C). This technique includes four stages freezing, thawing, and reculture. This freezing temperature inactivates the cell and thus can be preserved for a longer time. Any tissue from a plant can be used for cryopreservation, for example, meristems, embryos, endosperms, ovules, seeds, cultured plant cells, protoplasts, and calli. To prevent the damage caused to the cells (by freezing or thawing) various compounds like dimethylsulfoxide, glycerol, ethylene, propylene, sucrose, mannose, glucose, praline, acetamide, etc. are added during cryopreservation. These compounds are called cryoprotectants, which prevents the damage caused to cells by reducing the freezing point and super cooling point of water.

Cold Storage

This is a slow growth conservation method that conserves the germplasm at a low and nonfreezing temperature (1–9 °C). The main advantage of this method over cryopreservation is that it slows down growth of the plant material at cold storage (1–9 °C) in contrast to complete stoppage during cryopreservation, hence it protects the plants against cryogenic injuries. Moreover this method is simple, cost effective, and yields germplasms with better survival rates. Many outstanding examples of cold storage have recently been reported, for example, virus-free strawberry plants could be preserved at 10 °C for about 6 years and several grape plants have been stored for over 15 years by using cold storage at temperature around 9 °C (by transferring them in the fresh medium every year).

Low Pressure and Low Oxygen Storage

In this method atmospheric pressure (below 50 mmHg reduces plant tissue growth) and oxygen concentration surrounded by plant material are reduced, which causes the reduced availability of O_2, reduced production of CO_2, and hence photosynthetic activity is reduced, which inhibits plant tissue growth and dimension. These conditions may lead to slow and reduced growth of the plant material, which assists in increasing the shelf-life of many fruits, vegetables, and flowers. Therefore, conservation of germplasm by conventional methods has several limitations such as seed dormancy, short-lived seeds, seed-borne diseases, and high inputs of cost and labor. These modern techniques, like cryopreservation (freezing cells and tissues at -196 °C) and cold storage, help to overcome these problems.

Production of Secondary Metabolites

The process of in vitro culture of cells for the large scale production of secondary metabolites is complex, and involves the following aspects:

1. Selection of cell lines for high yield of secondary metabolites.

2. Large scale cultivation of plant cells.

3. Medium composition and effect of nutrients.

4. Elicitor-induced production of secondary metabolites.

5. Effect of environmental factors.

6. Biotransformation using plant cell cultures.

Selection of Cell Lines for High Yield of Secondary Metabolites

The very purpose of tissue culture is to produce high amounts of secondary metabolites. However, in general, majority of callus and suspension cultures produce less quantities of secondary metabolites. This is mainly due to the lack of fully differentiated cells in the cultures.

Some special techniques have been devised to select cell lines that can produce higher amounts of desired metabolites. These methods are ultimately useful for the separation of producer cells from the non-producer cells. The techniques commonly employed for cell line selection are cell cloning, visual or chemical analysis and selection for resistance.

Cell Cloning

This is a simple procedure and involves the growth of single cells (taken from a suspension cultures) in a suitable medium. Each cell population is then screened for the secondary metabolite formation. And only those cells with high-yielding ability are selected and maintained by sub-cloning.

Single Cell Cloning

There are certain practical difficulties in the isolation and culture of single cells.

Cell Aggregate Cloning

Compared to single cell cloning, cell aggregate cloning is much easier, hence preferred by many workers. A schematic representation of cell aggregate cloning for the selection of cells yielding high quantities of secondary metabolites is given in figure. A high yielding plant of the desired metabolite is selected and its explants are first cultured on a solid medium. After establishing the callus cultures, high metabolite producing calluses are identified, and they are grown in suspension cultures.

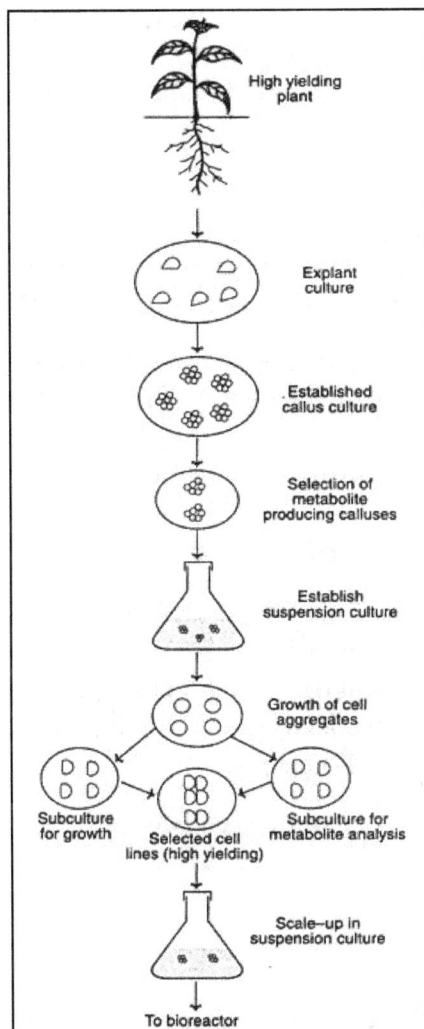

A schematic representation of cell aggregate cloning
for the selection of high yielding cell lines.

Cell aggregates from these cultures are grown on solid medium. The freshly developed cell aggregates (calluses) are divided into two parts. One half is grown further, while the other half is used for the quantitative analysis of the desired metabolite produced. The cell lines with high yield of secondary metabolites are selected and used for scale-up in suspension cultures. This is followed by large scale tissue culture in a bioreactor.

Visual or Chemical Analysis

A direct measurement of some of the secondary metabolites produced by cell lines can be done either by visual or chemical analysis. Visual identification of cell lines producing coloured secondary metabolites (pigments e.g., β-carotene, shikonin) will help in the selection of high-yielding cells. This method is quite simple and non-destructive. The major limitation is that the desired metabolite should be coloured.

Certain secondary metabolites emit fluorescence under UV light, and the corresponding clones can be identified. Some workers use simple, sensitive and inexpensive chemical analytical methods for quantitative estimation of desired metabolites. Analysis is carried out in some colonies derived from single cell cultures. Radioimmunoassay is the most commonly used analytical method. Micro spectrophotometry and fluorescent antibody techniques are also in use.

Selection for Resistance

Certain cells resistant to toxic compounds may lead to the formation of mutant cells which can overproduce a primary metabolite, and then a secondary metabolite. Such mutants can be selected and used to produce the desired metabolite in large quantities. One example is described.

Cell lines selected for resistance of 5-methyl- tryptophan (analogue of tryptophan) produce strains which can overproduce tryptophan. These tryptophan overproducing strains can synthesize 10-50 times higher levels of the natural auxin namely indole acetic acid (Note: The secondary metabolite indole acetic acid is derived from the primary metabolite tryptophan).

Large Scale (Mass) Cultivation of Plant Cells

In order to achieve industrial production of the desired metabolite, large scale cultivation of plant cells is required. Plant cells (20-150 μm in diameter) are generally 10-100 times larger than bacterial or fungal cell. When cultured, plant cells exhibit changes in volumes and thus variable shapes and sizes. Further, cultured cells have low growth rate and genetic instability. All these aspects have to be considered for mass cultivation of cells.

The following four different culture systems are widely used:

1. Free-cell suspension culture.

2. Immobilized cell culture.

3. Two-phase system culture.

4. Hairy root culture.

Free-cell Suspension Culture

Mass cultivation of plant cells is most frequently carried out by cell suspension cultures. Care should be taken to achieve good growth rate of cells and efficient formation of the desired secondary metabolite. Many specially designed bioreactors are in use for free-cell suspension cultures.

Some of these are listed below:

1. Batch bioreactors.

2. Continuous bioreactors.

3. Multistage bioreactors.

4. Airlift bioreactors.

5. Stirred tank bioreactors.

Two important aspects have to be considered for good success of suspension cultures:

1. Adequate and continuous oxygen supply.
2. Minimal generation of hydrodynamic stresses due to aeration agitation.

Immobilized Cell Cultures

Plant cells can be made immobile or immovable and used in culture systems. The cells are physically immobilized by entrapment. Besides individual cells, it is also possible to immobilize aggregate cells or even calluses. Homogenous suspensions of cells are most suitable for immobilization.

Surface immobilized plant cell (SIPC) technique efficiently retains the cells and allows them to grow at a higher rate. Further, through immobilization, there is better cell-to-cell contact, and the cells are protected from high liquid shear stresses. All this helps in the maximal production the secondary metabolite.

The common methods adopted for entrapment of cells are briefly described:

Entrapment of Cells in Gels

The cells or the protoplasts can be entrapped in several gels e.g., alginate, agar, agarose, carrageenin. The gels may be used either individually or in combination. The techniques employed for the immobilization of plant cells are comparable to those used for immobilization of microorganisms or other cells.

Entrapment of Cells in Nets or Foams

Polyurethane foams or nets with various pore sizes are used. The actively growing plant cells in suspension can be immobilized on these foams. The cells divide within the compartments of foam and form aggregates.

Entrapment of Cells in Hollow-fibre Membranes

Tubular hollow fibres composed of cellulose acetate silicone polycarbonate and organized into parallel bundles are used for immobilization of cells. It is possible to entrap cells within and between the fibres. Membrane entrapment is mechanically stable. However, it is more expensive than gel or foam immobilization.

Bioreactors for use of Immobilized Cells

Fluidized bed or fixed bed bioreactors are employed to use immobilized cells for large scale cultivation. In the fluidized-bed reactors, the immobilized cells are agitated by a flow of air or by pumping the medium. In contrast, in the fixed-bed bioreactor, the immobilized cells are held stationary (not agitated) and perfused at a slow rate with an aerated culture medium.

Biochemicals Produced by using Immobilized Cells

A selected list of the immobilized cells from selected plants and their utility to produce important bio-chemicals is given in table.

Table: A selected list of plant species with Immobilized cells employed for the production secondary metabolite(s).

Plant culture species	Immobilization Method	Substrate	Product
Catharanthus roseus	Entrapment in agarose	Cathenamine	Ajmalicine
Digitaiis Ianata	Entrapment in alginate	Digitoxin	Digoxin
Capsicum frutescens	Entrapment in polyurethane foam	Sucrose	Capsaicin
Catherranthus roseus	Entrapment in alginate, agarose, carrageenan	Sucrose	Ajmalicine
Petunia hybrid	Entrapment in hallow fibres	Sucrose	Phenolics
Morinda citifolia	Entrapment in alginate	Sucrose	Anthraquinone
Solanum aviculare	Attachment Polyphenylene beads	Sucrose	Steroid glycosides
Glycine max	Entrapment in hollow fibre	Sucrose	phenolics

Two-phase System Culture

Plant cells can be cultivated in an aqueous two phase system for the production of secondary metabolites. In this technique, the cells are kept apart from the product by separation in the bioreactor. This is advantageous since the product can be removed continuously. Certain polymers (e.g., dextran and polyethylene glycol for the separation of phenolic compounds) are used for the separation of phases.

Hairy Root Culture

Hairy root cultures are used for the production of root-associated metabolites. In general, these cultures have high growth rate and genetic stability. For the production of hairy root cultures, the explant material (plant tissue) is inoculated with the cells of the pathogenic bacterium, Agrobacterium rhizogenes. This organism contains root-inducing (Ri) plasmid that causes genetic transformation of plant tissues, which finally results in hairy root cultures. Hairy roots produced by plant tissues have metabolite features similar to that of normal roots.

Hairy root cultures are most recent organ culture systems and are successfully used for the commercial production of secondary metabolites. A selected list of the plants employed in hairy root cultures and the secondary metabolites produced is given in table.

Table: A selected list of plant species used in hairy root cultures for the production of secondary metabolite(s).

Plant species	Secondary metabolite(s)
Nicotiana tabacum	Nicotine, anatabine
Atropa belladonna	Atropine
Datura stramonium	Hyoscyamine
Lithospermum erythrorhizon	Shikonin
Catharanthus roseus	Ajmalicine, serpentine

Cinchona ledgeriana	Quinine alkaloids
Mentha Vulgaris	Monoterpenes
Solanum laciniatum	Steroid alkaloids

Medium Composition and Effects of Nutrients

The in vitro growth of the plant cells occurs in a suitable medium containing all the requisite elements. The ingredients of the medium effect the growth and metabolism of cells. For optimal production of secondary metabolites, a two-medium approach is desirable.

The first medium is required for good growth of cells (biomass growth) while the second medium, referred to as production medium promotes secondary metabolite formation. The effect of nutrients (carbon and nitrogen sources, phosphate, growth regulators, precursors, vitamins, metal ions) on different species in relation to metabolite formation are variable, some of them are briefly described.

Effects of Carbon Source

Carbohydrates influence the production of phytochemicals.

Some examples are given below:

1. Increase in sucrose concentration (in the range 4-10%) increases alkaloid production in Catharanthus roseus cultures.

2. Sucrose is a better carbon source than fructose or galactose for diosgenin production by Dioscorea deltoidea or Dalanites aegyptiaca cultures.

3. Low concentration of sucrose increases the production of ubiquinone-10 in tobacco cell cultures.

Effects of Nitrogen Source

The standard culture media usually contain a mixture of nitrate and ammonia as nitrogen source. Majority of plant cells can tolerate high levels of ammonia. The cultured cells utilize nitrogen for the biosynthesis of amino acids, proteins (including enzymes) and nucleic acids. The nitrogen containing primary metabolites directly influence the secondary metabolites.

In general, high ammonium ion concentrations inhibit secondary metabolite formation while lowering of ammonium nitrogen increases. It is reported that addition of KNO_3 and NH_4NO_3 inhibited anthocyanin (by 90%) and alkaloid (by 80%) production.

Effects of Phosphate

Inorganic phosphate is essential for photosynthesis and respiration (glycolysis). In addition, many secondary metabolites are produced through phosphorylated intermediates, which subsequently release the phosphate e.g., phenylpropanoids, terpenes, terpenoids. In general, high phosphate levels promote cell growth and primary metabolism while low phosphate concentrations are beneficial for secondary product formation. However, this is not always correct.

Increase in phosphate concentration in the medium may increase, decrease or may not affect product formation e.g.:

1. Increased phosphate concentration increases alkaloid (in Catharanthus roseus), anthraquinone (in Morinda citrifolia) and diosgenin (in Dioscorea deltoidea) production.

2. Decreased phosphate level in the medium increases the formation of anthocyanins and phenolics (in Catharanthus roseus), alkaloids (in Peganum harmala) and solasodine (in Solanum lanciatum).

3. Phosphate concentration (increase or decrease) has no effect on protoberberine (an alkaloid) production by Berberis sp.

Effects of Plant Growth Regulators

Plant growth regulators (auxins, cytokinins) influence growth, metabolism and differentiation of cultured cells. There are a large number of reports on the influence of growth regulators for the production of secondary metabolites in cultured cells. A few examples are given below:

1. Addition of auxins (indole acetic acid, indole pyruvic acid, naphthalene acetic acid) enhanced the production of diosgenin in the cultures of Balanites aegyptiaca.

2. Auxins may inhibit the production of certain secondary metabolites e.g., naphthalene acetic acid and indole acetic acid inhibited the synthesis of anthocyanin in carrot cultures.

3. Another auxin, 2, 4-dichlorophenoxy acetate (2, 4-D) inhibits the production of alkaloids in the cultures of tobacco, and shikonin formation in the cultures of Lithospermum erythrorhizon.

4. Cytokinins promote the production of secondary metabolites in many tissue cultures e.g., ajmalicine in Catharanthus roseus; scopolin and scopoletin in tobacco; carotene in Ricinus sp.

5. In some tissue cultures, cytokinins inhibit product formation e.g., anthroquinones in Morinda citrifolia; shikonin in Lithospermum erythroshizon; nicotine in tobacco.

In actual practice, a combination of auxins and cytokinins is used to achieve maximum production of secondary metabolites in culture systems.

Effects of Precursors

The substrate molecules that are incorporated into the secondary metabolites are referred to as precursors. In general, addition of precursors to the medium enhances product formation, although they usually inhibit the growth of the culture e.g., alkaloid synthesis in Datura cultures in increased while growth is inhibited by the addition of ornithine, phenylalanine, tyrosine and sodium phenyl pyruvate; precursors tryptamine and secologanin increase ajmalicine production in C. roseus cultures.

Elicitor-induced Production of Secondary Metabolites

The production of secondary metabolites in plant cultures is generally low and does not meet the commercial demands. There are continuous efforts to understand the mechanism of product formation at the molecular level, and exploit for increased production. The synthesis of majority of secondary metabolites involves multistep reactions and many enzymes. It is possible to stimulate any step to increase product formation.

Elicitors are the compounds of biological origin which stimulate the production of secondary metabolites, and the phenomenon of such stimulation is referred to as elicitation. Elicitors produced within the plant cells are endogenous elicitors e.g., pectin, pectic acid, cellulose, other polysaccharides. When the elicitors are produced by the microorganisms, they are referred to as exogenous elicitors e.g., chitin, chitosan, glucans. All the elicitors of biological origin are biotic elicitors.

The term abiotic elicitors is used to represent the physical (cold, heat, UV light, osmotic pressure) and chemical agents (ethylene, fungicides, antibiotics, salts of heavy metals) that can also increase the product formation. However, the term abiotic stress is used for abiotic elicitors, while elicitors exclusively represent biological compounds.

Phytoalexins

Plants are capable of defending themselves when attacked by microorganisms, by producing antimicrobial compounds collectively referred to as phytoalexins. Phytoalexins are the chemical weapons of defense against pathogenic microorganisms. Some of the phytoalexins that induce the production of secondary metabolites are regarded as elicitors. Some chemicals can also act as elicitors e.g., actinomycin-D, sodium salt of arachidonic acid, ribonuclease-A, chitosan, poly-L- lysine, nigeran. These compounds are regarded as chemically defined elicitors.

Interactions for Elicitor Formation

Elicitors are compounds involved in plant- microbe interaction. Three different types of interactions between plants and microorganisms are known that lead to the formation of elicitors.

1. Direct release of elicitor by the microorganisms.

2. Microbial enzymes that can act as elicitors. e.g. endopolygalacturonic acid lyase from Erwinia carotovara.

3. Release of phytoalexins by the action of plant enzymes on cell walls of microorganisms which in turn stimulate formation elicitors from plant cell walls e.g., chitosan from Fusarium cell walls; α-1, 3-endoglucanase from Phytophthora cell walls.

Methodology of Elicitation

Selection of Microorganisms

A wide range of microorganisms (viruses, bacteria, algae and fungi) that need not be pathogens have been tried in cultures for elicitor induced production of secondary metabolites. Based on the favourable elicitor response, an ideal microorganism is selected. The quantity of the microbial inoculum is important for the formation elicitor.

Co-culture

Plant cultures (frequently suspension cultures) are inoculated with the selected microorganism to form co-cultures. The cultures are transferred to a fresh medium prior to the inoculation with microorganism. This helps to stimulate the secondary metabolism.

Co-cultures of plant cells with microorganisms may sometimes have inhibitory effect on the plant cells. In such a case, elicitor preparations can be obtained by culturing the selected microorganism on a tissue culture medium, followed by homogenization and autoclaving of the entire culture. This process releases elicitors. In case of heat labile elicitors, the culture homogenate has to be filter sterilized (instead of autoclaving).

In some co-culture systems, direct contact of plant cells and microorganisms can be prevented by immobilization (entrapment) of one of them. In these cultures, plant microbial interaction occurs by diffusion of the elicitor compounds through the medium.

Mechanism of Action of Elicitors

Elicitors are found to activate genes and increase the synthesis of mRNAs encoding enzymes responsible for the ultimate biosynthesis secondary metabolites. There are some recent reports suggesting the involvement of elicitor mediated calcium-based signal transduction systems that promotes the product formation. When the cells are pretreated with a calcium chelate (EDTA) prior to the addition of elicitor, there occurs a decrease in the production of secondary metabolite.

Elicitor-induced Products in Cultures

In table a selected list of elicitor-induced secondary metabolites produced in culture systems is given.

Table: A selected list of elicitor-induced secondary metabolite production in plant cell cultures.

Elicitor microorganism	Plant cell culture(s)	Secondary metabolite(s)
Aspergillus niger	Cinchona ledgeriana, Rubia tinctoria	Anthraquinones
Pythium aphanidermatum	Catharanthus roseus	Ajmalicine, Strictosidine, Catharanthine
Borytis sp	Papaver somniferum	Sanguinarine
Phytophthora Megasperma	Glycine max	Isoflavonoids Gluceollin
Dendryphion sp	Papaver sominferum	Sanguinarine
Alternaria sp	Phaseolus vulgaris	Phaseollin
Fusarium sp	Apium graveolens	Furanocoumarins
Phythium aphanidermatum	Daucus carota	Anthocynins
Penicillium	Sanguinaria	Benzophenan-
Expansum	Canadensis	Thridine Alkaloids

Effects of Environmental Factors

The physical factors namely light, incubation temperature, pH of the medium and aeration of cultures influence the production of secondary metabolites in cultures.

Effects of Light

Light is absolutely essential for the carbon fixation (photosynthesis) of field-grown plants. Since the carbon fixation is almost absent or very low in plant tissue cultures, light has no effect on the primary metabolism.

However, the light-mediated enzymatic reactions indirectly influence the secondary metabolite formation. The quality of light is also important. Some examples of light- stimulated product formations are given:

1. Blue light enhances anthocyanin production in Haplopappus gracilis cell suspensions.

2. White light increases the formation of anthocyanin in the cultures of Catharanthus roseus, Daucus carota and Helianthus tuberosus.

3. White or blue light inhibits naphthoquinone biosynthesis in callus cultures of Lithospermum erythrorhizon.

Effects of Incubation Temperature

The growth of cultured cells is increased with increase in temperature up to an optimal temperature (25-30 °C). However, at least for the production some secondary metabolites lower temperature is advantageous. For instance, in C. roseus cultures, indole alkaloid production is increased by two fold when incubated at 16 °C instead of 27 °C. Increased temperature was also found to reduce the production of caffeine (by C. sineneis) and nicotine (by N. tabacum).

Effect of pH of the Medium

For good growth of cultures, the pH of the medium is in the range of 5 to 6. There are reports indicating that pH of the medium influences the formation of secondary metabolites. e.g., production of anthocyanin by cultures of Daucus carota was much less when incubated at pH 5.5 than at pH 4.5. This is attributed to the increased degradation of anthocyanin at higher pH.

Aeration of Cultures

Continuous aeration is needed for good growth of cultures, and also for the efficient production of secondary metabolites.

Biotransformation using Plant Cell Cultures

The conversion of one chemical into another (i.e., a substrate into a final product) by using biological systems (i.e. cell suspensions) as biocatalysts is regarded as biotransformation or bioconversion. The biocatalyst may be free or immobilized, and the process of biotransformation may involve one or more enzymes.

The biotechnological application of plant cell cultures in biotransformation reactions involves the conversion of some less important substances to valuable medicinal or commercially important products. In biotransformation, it is necessary to select such cell lines that possess the enzymes for catalysing the desired reactions. Bioconversions may involve many types of reactions e.g., hydroxylation, reduction, glycosylation.

A good example of biotransformation by plant cell cultures is the large scale production of cardiovascular drug digoxin from digitoxin by Digitali lanata. Digoxin production is carried out by immobilized cells of D. lanata in airlift bioreactors. Cell cultures of Digitalis purpurea or Stevia rebaudiana can convert steviol into steviobiocide and steviocide which are 100 times sweeter than cane sugar.

A selected list of biotransformation's carried out in plant cell cultures is given in table.

Table: A selected list of elicitor-induced secondary metabolite production in plant cell cultures.

Elicitor microorganism	Plant cell culture(s)	Secondary metabolite(s)
Aspergillus niger	Cinchona ledgeriana, Rubia tinctoria	Anthrpuinones
Pythium aphanidermatum	Catharanthus roseus	Ajmalicine, Strictosidine, Catharanthine
Botrytis sp	Papaver somniferum	Sanguinarine
Phytophtora megasperma	Glycine max	Isoflavonoids Gluceolin
Dendryphion sp	Papaver somniferum	Sanguinarine
Alternaria sp	Phaseolus vulgaris	Phaseollin
Fusarium sp	Apium graveolens	Furanocoumarins
Phythium aphanidermatum	Daucus carota	Anthocynins
Penicillium	Sanguinaria	Benzophenan-
Expansum	canadensis	Thridine Alkaloids

Secondary Metabolite Release and Analysis

The methods employed for the separation and purification of secondary metabolites from cell cultures are the same as that used for plants.

Sometimes, the products formed within the cells are released into the medium, making the isolation and analysis easy. For the secondary metabolites stored within the vacuoles of cells, two membranes (plasma membrane and tonoplast) have to be disrupted. Permeabilizing agents such as dimethyl sulfoxide (DMSO) can be used for the release of products.

In general, separation and purification of products from plant cell cultures are expensive, therefore every effort is made to make them cost-effective. Two approaches are made in this direction:

1. Production of secondary metabolite should be as high as possible.

2. Formation of side product(s) which interfere with separation must be made minimal.

Once a good quantity of the product is released into the medium, separation and purification techniques (e.g. extraction) can be used for its recovery. These techniques largely depend on the nature of the secondary metabolite.

References

- Propagation-by-Tissue-Culture, Biotechnology-Plant: agritech.tnau.ac.in, Retrieved 18 April, 2019

- Types-of-tissue-culture-biotechnology, bioinformatics: biotechnologynotes.com, Retrieved 14 June, 2019

- Plant-tissue-culture-media, recent-advances-in-plant-in-vitro-culture: intechopen.com, Retrieved 28 February, 2019

- Tools-and-techniques-used-for-plant-tissue-culture, plant-tissue-culture, plants: biologydiscussion.com, Retrieved 19 May, 2019

- Totipotency-meaning-expression-and-importance-plant-tissue-culture, totipotency, plant-tissues: biologydiscussion.com, Retrieved 28 April, 2019

- Organogenesis, agricultural-and-biological-sciences: sciencedirect.com, Retrieved 1 August, 2019

- Organogenesis, definition-and-factors, influencing-organogenesis-plant-tissue-culture, tissue-culture, botany: biologydiscussion.com, Retrieved 13 July, 2019

- Somatic-embryogenesis, meaning, history, principles, protocols-and-importance, plant-tissues: biologydiscussion.com, Retrieved 29 April, 2019

- Meristem-culture: biologyreader.com, Retrieved 30 March, 2019

- The-range-of-regenerative-capability, regeneration-biology, science: britannica.com, Retrieved 5 january, 2019

- Production-of-haploid-plants, haploid-plants, plants: biologydiscussion.com, Retrieved 17 February, 2019

- Applications-of-haploid-plants, haploid-plants, plants: biologydiscussion.com, Retrieved 24 March, 2019

- Germplasm-conservation, agricultural-and-biological-sciences: sciencedirect.com, Retrieved 21 March, 2019

- Secondary-metabolites-in-plant-cultures-applications-and-production, plant-biotechnology, biotechnology: biologydiscussion.com, Retrieved 12 April, 2019

Chapter 3

Plant Breeding

The science of changing the traits of plants in order to produce desired characteristics is known as plant breeding. It is primarily used to improve the quality of nutrition for humans and animals. It is also involved in making the plants disease and drought resistant. The following chapter elucidates these varied processes and mechanisms associated with the breeding of plants.

Plant breeding is the genetic improvement of the crop in order to create desired plant types that are better suited for cultivation, give better yields and are disease resistant. Conventional plant breeding is in practice from 9,000-11,000 years ago. Most of our major food crops are derived from the domesticated varieties.

But now due to advancements in genetics, molecular biology and tissue culture, plant breeding is being carried out by using molecular genetics tools. Classical plant breeding includes hybridization (crossing) of pure lines, artificial selection to produce plants with desirable characters of higher yield, nutrition and resistance to diseases.

When the breeders wish to incorporate desired characters (traits) into the crop plants, they should increase yield and improve the quality. Increased tolerance to salinity, extreme temperatures, drought, resistance to viruses, fungi, bacteria and increased tolerance to insect pests should also be the desired traits in these crop plants.

Goals

The plant breeder usually has in mind an ideal plant that combines a maximum number of desirable characteristics. These characteristics may include resistance to diseases and insects; tolerance to heat, soil salinity, or frost; appropriate size, shape, and time to maturity; and many other general and specific traits that contribute to improved adaptation to the environment, ease in growing and handling, greater yield, and better quality. The breeder of horticultural plants must also consider aesthetic appeal. Thus the breeder can rarely focus attention on any one characteristic but must take into account the manifold traits that make the plant more useful in fulfilling the purpose for which it is grown. Plant breeding is an important tool in promoting global food security, and many staple crops have been bred to better withstand extreme weather conditions associated with global warming, such as drought or heat waves.

Increase of Yield

One of the aims of virtually every breeding project is to increase yield. This can often be brought about by selecting obvious morphological variants. One example is the selection of dwarf, early maturing varieties of rice. These dwarf varieties are sturdy and give a greater yield of grain. Furthermore, their early maturity frees the land quickly, often allowing an additional planting of rice or other crop the same year.

Another way of increasing yield is to develop varieties resistant to diseases and insects. In many cases the development of resistant varieties has been the only practical method of pest control. Perhaps the most important feature of resistant varieties is the stabilizing effect they have on production and hence on steady food supplies. Varieties tolerant to drought, heat, or cold provide the same benefit.

Modifications of Range and Constitution

Another common goal of plant breeding is to extend the area of production of a crop species. A good example is the modification of grain sorghum since its introduction to the United States in the 1750s. Of tropical origin, grain sorghum was largely confined to the southern Plains area and the Southwest, but earlier-maturing varieties were developed, and grain sorghum is now an important crop as far north as North Dakota.

Development of crop varieties suitable for mechanized agriculture has become a major goal of plant breeding in recent years. Uniformity of plant characters is very important in mechanized agriculture because field operations are much easier when the individuals of a variety are similar in time of germination, growth rate, size of fruit, and so on. Uniformity in maturity is, of course, essential when crops such as tomatoes and peas are harvested mechanically.

The nutritional quality of plants can be greatly improved by breeding. For example, it is possible to breed varieties of corn (maize) much higher in lysine than previously existing varieties. Breeding high-lysine maize varieties for those areas of the world where maize is the major source of this nutritionally essential amino acid has become a major goal in plant breeding. This "biofortification" of food crops, a term which also includes genetic modification, has been shown to improve nutrition and is especially useful in developing areas where nutritional deficiencies are common and medical infrastructure may be lacking.

In breeding ornamental plants, attention is paid to such factors as longer blooming periods, improved keeping qualities of flowers, general thriftiness, and other features that contribute to usefulness and aesthetic appeal. Novelty itself is often a virtue in ornamentals, and the spectacular, even the bizarre, is often sought.

Evaluation of Plants

The appraisal of the value of plants so that the breeder can decide which individuals should be discarded and which allowed to produce the next generation is a much more difficult task with some traits than with others.

Qualitative Characters

The easiest characters, or traits, to deal with are those involving discontinuous, or qualitative, differences that are governed by one or a few major genes. Many such inherited differences exist, and they frequently have profound effects on plant value and utilization. Examples are starchy versus sugary kernels (characteristic of field and sweet corn, respectively) and determinant versus indeterminant habit of growth in green beans (determinant varieties are adapted to mechanical harvesting). Such differences can be seen easily and evaluated quickly, and the expression of the traits remains the same regardless of the environment in which the plant grows. Traits of this type are termed highly heritable.

Quantitative Characters

In other cases, however, plant traits grade gradually from one extreme to another in a continuous series, and classification into discrete classes is not possible. Such variability is termed quantitative. Many traits of economic importance are of this type; e.g., height, cold and drought tolerance, time to maturity, and, in particular, yield. These traits are governed by many genes, each having a small effect. Although the distinction between the two types of traits is not absolute, it is nevertheless convenient to designate qualitative characters as those involving discrete differences and quantitative characters as those involving a graded series.

Quantitative characters are much more difficult for the breeder to control, for three main reasons:

1. the sheer numbers of the genes involved make hereditary change slow and difficult to assess;

2. the variations of the traits involved are generally detectable only through measurement and exacting statistical analyses; and

3. most of the variations are due to the environment rather than to genetic endowment; for example, the heritability of certain traits is less than 5 percent, meaning that 5 percent of the observed variation is caused by genes and 95 percent is caused by environmental influences.

It follows that carefully designed experiments are required to distinguish plants that are superior because they carry desirable genes from those that are superior because they happen to grow in a favourable site.

Methods of Plant Breeding

Mating Systems

Angiosperm mating systems devolve about the type of pollination, or transferal of pollen from flower to flower. A flower is self-pollinated (a "selfer") if pollen is transferred to it from any flower of the same plant and cross-pollinated (an "outcrosser" or "outbreeder") if the pollen comes from a flower on a different plant. About half of the more important cultivated plants are naturally cross-pollinated, and their reproductive systems include various devices that encourage cross-pollination—e.g., protandry (pollen shed before the ovules are mature, as in the carrot and walnut), dioecy (male and female parts are borne on different plants, as in the date palm, asparagus, and hops), and genetically determined self-incompatibility (inability of pollen to grow on the stigma of the same plant, as in white clover, cabbage, and many other species).

Other plant species, including a high proportion of the most important cultivated plants such as wheat, barley, rice, peas, beans, and tomatoes, are predominantly self-pollinating. There are relatively few reproductive mechanisms that promote self-pollination; the most positive of which is failure of the flowers to open (cleistogamy), as in certain violets. In barley, wheat, and lettuce the pollen is shed before or just as the flowers open, and in the tomato pollination follows opening of the flower, but the stamens form a cone around the stigma. In such species there is always a risk of unwanted cross-pollination.

In controlled breeding procedures it is imperative that pollen from the desired male parent, and no other pollen, reaches the stigma of the female parent. When stamens and pistils occur in the same flower, the anthers must be removed from flowers selected as females before pollen is shed. This is usually done with forceps or scissors. Protection must also be provided from "foreign" pollen. The most common method is to cover the flower with a plastic or paper bag. When the stigma of the female parent becomes receptive, pollen from the desired male parent is transferred to it, often by breaking an anther over the stigma, and the protective bag is replaced. The production of certain hybrids is, therefore, tedious and expensive because it often requires a series of delicate, exacting, and properly timed hand operations. When male and female parts occur in separate flowers, as in corn (maize), controlled breeding is easier.

Hand pollination: Apple blossoms being hand-pollinated.

A cross-pollinated plant, which has two parents, each of which is likely to differ in many genes, produces a diverse population of plants hybrid (heterozygous) for many traits. A self-pollinated plant, which has only one parent, produces a more uniform population of plants pure breeding (homozygous) for many traits.

Breeding Self-pollinated Species

The breeding methods that have proved successful with self-pollinated species are:

1. mass selection;

2. pure-line selection;

3. hybridization, with the segregating generations handled by the pedigree method, the bulk method, or by the backcross method; and

4. development of hybrid varieties.

Mass Selection

In mass selection, seeds are collected from (usually a few dozen to a few hundred) desirable appearing individuals in a population, and the next generation is sown from the stock of mixed seed. This procedure, sometimes referred to as phenotypic selection, is based on how each individual looks.

Mass selection has been widely used to improve old "land" varieties—varieties that have been passed down from one generation of farmers to the next over long periods—and is common in horticulture.

An alternative approach that has no doubt been practiced for thousands of years is simply to eliminate undesirable types by destroying them in the field. The results are similar whether superior plants are saved or inferior plants are eliminated: seeds of the better plants become the planting stock for the next season.

A modern refinement of mass selection is to harvest the best plants separately and to grow and compare their progenies. The poorer progenies are destroyed and the seeds of the remainder are harvested. Selection is now based not solely on the appearance of the parent plants but also on the appearance and performance of their progeny. Progeny selection is usually more effective than phenotypic selection when dealing with quantitative characters of low heritability. progeny testing requires an extra generation; hence gain per cycle of selection must be double that of simple phenotypic selection to achieve the same rate of gain per unit time.

Mass selection, with or without progeny test, is perhaps the simplest and least expensive of plant-breeding procedures. It finds wide use in the breeding of certain forage species, which are not important enough economically to justify more detailed attention.

Pure-line Selection

Pure-line selection generally involves three more or less distinct steps:

1. numerous superior appearing plants are selected from a genetically variable population;

2. progenies of the individual plant selections are grown and evaluated by simple observation, frequently over a period of several years; and

3. when selection can no longer be made on the basis of observation alone, extensive trials are undertaken, involving careful measurements to determine whether the remaining selections are superior in yielding ability and other aspects of performance. Any progeny superior to an existing variety is then released as a new "pure-line" variety. Much of the success of this method during the early 1900s depended on the existence of genetically variable land varieties that were waiting to be exploited. They provided a rich source of superior pure-line varieties, some of which are still represented among commercial varieties. In recent years the pure-line method as outlined above has decreased in importance in the breeding of major cultivated species; however, the method is still widely used with the less important species that have not yet been heavily selected.

A variation of the pure-line selection method that dates back centuries is the selection of single-chance variants, mutations or "sports" in the original variety. A very large number of varieties that differ from the original strain in characteristics such as colour, lack of thorns or barbs, dwarfness, and disease resistance have originated in this fashion.

Hybridization

The object of hybridization is to combine desirable genes found in two or more different varieties and to produce pure-breeding progeny superior in many respects to the parental types.

Genes, however, are always in the company of other genes in a collection called a genotype. The plant breeder's problem is largely one of efficiently managing the enormous numbers of genotypes that occur in the generations following hybridization. As an example of the power of hybridization in creating variability, a cross between hypothetical wheat varieties differing by only 21 genes is capable of producing more than 10,000,000,000 different genotypes in the second generation. While the great majority of these second generation genotypes are hybrid (heterozygous) for one or more traits, it is statistically possible that 2,097,152 different pure-breeding (homozygous) genotypes can occur, each potentially a new pure-line variety. These numbers illustrate the importance of efficient techniques in managing hybrid populations, for which purpose the pedigree procedure is most widely used.

Pedigree breeding starts with the crossing of two genotypes, each of which have one or more desirable characters lacked by the other. If the two original parents do not provide all of the desired characters, a third parent can be included by crossing it to one of the hybrid progeny of the first generation (F_1). In the pedigree method superior types are selected in successive generations, and a record is maintained of parent–progeny relationships.

The F_2 generation (progeny of the crossing of two F_1 individuals) affords the first opportunity for selection in pedigree programs. In this generation the emphasis is on the elimination of individuals carrying undesirable major genes. In the succeeding generations the hybrid condition gives way to pure breeding as a result of natural self-pollination, and families derived from different F_2 plants begin to display their unique character. Usually one or two superior plants are selected within each superior family in these generations. By the F_5 generation the pure-breeding condition (homozygosity) is extensive, and emphasis shifts almost entirely to selection between families. The pedigree record is useful in making these eliminations. At this stage each selected family is usually harvested in mass to obtain the larger amounts of seed needed to evaluate families for quantitative characters. This evaluation is usually carried out in plots grown under conditions that simulate commercial planting practice as closely as possible. When the number of families has been reduced to manageable proportions by visual selection, usually by the F_7 or F_8 generation, precise evaluation for performance and quality begins. The final evaluation of promising strains involves (1) observation, usually in a number of years and locations, to detect weaknesses that may not have appeared previously; (2) precise yield testing; and (3) quality testing.

The bulk-population method of breeding differs from the pedigree method primarily in the handling of generations following hybridization. The F_2 generation is sown at normal commercial planting rates in a large plot. At maturity the crop is harvested in mass, and the seeds are used to establish the next generation in a similar plot. No record of ancestry is kept. During the period of bulk propagation natural selection tends to eliminate plants having poor survival value. Two types of artificial selection also are often applied: (1) destruction of plants that carry undesirable major genes and (2) mass techniques such as harvesting when only part of the seeds are mature to select for early maturing plants or the use of screens to select for increased seed size. Single plant selections are then made and evaluated in the same way as in the pedigree method of breeding. The chief advantage of the bulk population method is that it allows the breeder to handle very large numbers of individuals inexpensively.

Often an outstanding variety can be improved by transferring to it some specific desirable character that it lacks. This can be accomplished by first crossing a plant of the superior variety to a plant of the donor variety, which carries the trait in question, and then mating the progeny back to a plant having the genotype of the superior parent. This process is called backcrossing. After five or six backcrosses the progeny will be hybrid for the character being transferred but like the superior parent for all other genes. Selfing the last backcross generation, coupled with selection, will give some progeny pure breeding for the genes being transferred. The advantages of the backcross method are its rapidity, the small number of plants required, and the predictability of the outcome. A serious disadvantage is that the procedure diminishes the occurrence of chance combinations of genes, which sometimes leads to striking improvements in performance.

Hybrid Varieties

The development of hybrid varieties differs from hybridization in that no attempt is made to produce a pure-breeding population; only the F_1 hybrid plants are sought. The F_1 hybrid of crosses between different genotypes is often much more vigorous than its parents. This hybrid vigour, or heterosis, can be manifested in many ways, including increased rate of growth, greater uniformity, earlier flowering, and increased yield, the last being of greatest importance in agriculture.

By far the greatest development of hybrid varieties has been in corn (maize), primarily because its male flowers (tassels) and female flowers (incipient ears) are separate and easy to handle, thus proving economical for the production of hybrid seed. The production of hand-produced F1 hybrid seed of other plants, including ornamental flowers, has been economical only because greenhouse growers and home gardeners have been willing to pay high prices for hybrid seed.

Recently, however, a built-in cellular system of pollination control has made hybrid varieties possible in a wide range of plants, including many that are self-pollinating, such as sorghums. This system, called cytoplasmic male sterility, or cytosterility, prevents normal maturation or function of the male sex organs (stamens) and results in defective pollen or none at all. It obviates the need for removing the stamens either by hand or by machine. Cytosterility depends on the interaction between male sterile genes (R + r) and factors found in the cytoplasm of the female sex cell. The genes are derived from each parent in the normal Mendelian fashion, but the cytoplasm (and its factors) is provided by the egg only; therefore, the inheritance of cytosterility is determined by the female parent. All plants with fertile cytoplasm produce viable pollen, as do plants with sterile cytoplasm but at least one R gene; plants with sterile cytoplasm and two r genes are male sterile (produce defective pollen).

The production of F_1 hybrid seed between two strains is accomplished by interplanting a sterile version of one strain (say A) in an isolated field with a fertile version of another strain (B). Since strain A produces no viable pollen, it will be pollinated by strain B, and all seeds produced on strain A plants must therefore be F_1 hybrids between the strains. The F_1 hybrid seeds are then planted to produce the commercial crop.

Breeding Cross-pollinated Species

The most important methods of breeding cross-pollinated species are (1) mass selection; (2) development of hybrid varieties; and (3) development of synthetic varieties. Since cross-pollinated species are naturally hybrid (heterozygous) for many traits and lose vigour as they become purebred (homozygous), a goal of each of these breeding methods is to preserve or restore heterozygosity.

Mass Selection

Mass selection in cross-pollinated species takes the same form as in self-pollinated species; i.e., a large number of superior appearing plants are selected and harvested in bulk and the seed used to produce the next generation. Mass selection has proved to be very effective in improving qualitative characters, and, applied over many generations, it is also capable of improving quantitative characters, including yield, despite the low heritability of such characters. Mass selection has long been a major method of breeding cross-pollinated species, especially in the economically less important species.

Hybrid Varieties

The outstanding example of the exploitation of hybrid vigour through the use of F_1 hybrid varieties has been with corn (maize). The production of a hybrid corn variety involves three steps: (1) the selection of superior plants; (2) selfing for several generations to produce a series of inbred lines, which although different from each other are each pure-breeding and highly uniform; and (3) crossing selected inbred lines. During the inbreeding process the vigour of the lines decreases drastically, usually to less than half that of field-pollinated varieties. Vigour is restored, however, when any two unrelated inbred lines are crossed, and in some cases the F_1 hybrids between inbred lines are much superior to open-pollinated varieties. An important consequence of the homozygosity of the inbred lines is that the hybrid between any two inbreds will always be the same. Once the inbreds that give the best hybrids have been identified, any desired amount of hybrid seed can be produced.

Pollination in corn (maize) is by wind, which blows pollen from the tassels to the styles (silks) that protrude from the tops of the ears. Thus controlled cross-pollination on a field scale can be accomplished economically by interplanting two or three rows of the seed parent inbred with one row of the pollinator inbred and detasselling the former before it sheds pollen. In practice most hybrid corn is produced from "double crosses," in which four inbred lines are first crossed in pairs (A × B and C × D) and then the two F_1 hybrids are crossed again (A × B) × (C × D). The double-cross procedure has the advantage that the commercial F_1 seed is produced on the highly productive single cross A × B rather than on a poor-yielding inbred, thus reducing seed costs. In recent years cytoplasmic male sterility, described earlier, has been used to eliminate detasselling of the seed parent, thus providing further economies in producing hybrid seed.

Much of the hybrid vigour exhibited by F_1 hybrid varieties is lost in the next generation. Consequently, seed from hybrid varieties is not used for planting stock but the farmer purchases new seed each year from seed companies.

Perhaps no other development in the biological sciences has had greater impact on increasing the quantity of food supplies available to the world's population than has the development of hybrid corn (maize). Hybrid varieties in other crops, made possible through the use of male sterility, have also been dramatically successful and it seems likely that use of hybrid varieties will continue to expand in the future.

Synthetic Varieties

A synthetic variety is developed by intercrossing a number of genotypes of known superior combining ability—i.e., genotypes that are known to give superior hybrid performance when crossed in all combinations. (By contrast, a variety developed by mass selection is made up of genotypes bulked together without having undergone preliminary testing to determine their performance in hybrid combination.) Synthetic varieties are known for their hybrid vigour and for their ability to produce usable seed for succeeding seasons. Because of these advantages, synthetic varieties have become increasingly favoured in the growing of many species, such as the forage crops, in which expense prohibits the development or use of hybrid varieties.

Distribution and Maintenance of New Varieties

The benefits of superior new varieties obviously cannot be realized until sufficient seed has been produced to permit commercial production. Although the primary function of the plant breeder is to develop new varieties, he usually also carries out an initial small-scale seed increase. Seed thus produced is called breeders seed. The next stage is the multiplication of breeders seed to produce foundation seed. Production of foundation seed is usually carried out by seed associations or institutes, whose work is regulated by government agencies. The third step is the production of certified seed, the progeny of foundation seed, produced on a large scale by specialized seed growers for general sale to farmers and gardeners. Certified seed must be produced and handled in such a way as to meet the standards set by the certifying agency (usually a seed association). Seed associations are also usually responsible for maintaining the purity of new varieties once they have been released for commercial production.

The distribution of new varieties developed by commercial plant-breeding companies is often through seed associations, but many reputable companies market their products without following the official certification process. In some countries, particularly in Europe, new varieties can be patented for periods up to 15 years or more, during which time the breeder has an exclusive right to reproduce and sell the variety.

Disease Resistance in Plants

The genetic engineering techniques for disease resistance are:

Developing Virus Resistance Food Crops

Viruses are among the most ubiquitous pests in agriculture. Scientists are working to develop viral resistance in a variety of crops including squash, potato, sweet potato, wheat, papaya and raspberries.

Viruses are studied widely because they not only cause disease in humans, plants, animals and insects, but also are used as tools in the study of molecular biology and, in some cases, in the development of vaccines to fight the diseases they can cause.

Several techniques for virus resistance have been developed. These include viral coat protein technology and multiple gene transfers. A viral coat protein acts like a vaccine, causing the plant to develop resistance to the particular virus. Transferring the gene for a viral coat protein, a part of the outer shell of a virus that does not cause disease, into a plant acts like a vaccine for the plant.

The plant is then able to resist the virus, analogous to the way vaccines keep us from getting certain diseases like measles. The advantage of introducing only the coat protein is that it induces resistance without the introduction of the actual virus. The technique has been used successfully in many plants against several different viruses.

The first genetically engineered virus-resistant food crop in the marketplace was yellow crookneck squash. Using the viral coat protein approach, this squash was engineered to resist the watermelon mosaic virus and the zucchini yellow mosaic virus. Potatoes are highly susceptible to many viruses, including the potato mosaic virus and the potato leaf roll virus.

A leaf roll virus epidemic in 1996 was responsible for heavy potato crop losses in Idaho. The virus, spread by aphids, damaged the potatoes to the point that they were unmarketable. Scientists in Mexico, in collaboration with researchers at Monsanto, have developed potatoes resistant to several forms of this virus. Research on disease-resistant potatoes is continuing at other laboratories.

The feathery mottle virus has a damaging effect on sweet potatoes. In 1991, researchers began genetically engineering varieties of sweet potato grown in Africa, where it is an important subsistence crop. The sweet potato was engineered with coat protein from this virus and replicase genes. Replicase is an enzyme involved in the duplication of certain viral RNA molecules.

Current field-testing has demonstrated successful gene transformations and the desired development of resistance to sweet potato feathery mottle virus. Although wheat is an important food source, development of genetically engineered varieties has been slower than in corn, soy and cotton.

A major pest in wheat is barley yellow dwarf virus, which can cause damage in major wheat-growing regions such as North Dakota, because no resistant strains are known. Work is in progress to engineer resistance to this disease using the viral coat protein technique.

The wheat genome is highly complex-ten to twenty times larger than that of cotton or rice-and carries an exceptionally large amount of repetitive DNA sequences. Thus, targeting particular genes is challenging, and transgenic wheat biotechnology has advanced more slowly than that of other crops.

The papaya crop in Hawaii was nearly wiped out in the 1950s by the papaya ring-spot virus (PRSV). Transmitted by aphids, this virus causes one of the most serious diseases of papaya worldwide. Work to develop a transgenic virus-resistant variety began in the late 1980s. By 1992, resistant lines were field-tested; approvals for commercialization were granted in 1997.

The transgenic- resistant papaya is now in wide use in Hawaii, and similar work is in progress in the Philippines, Malaysia, Thailand, Vietnam and Indonesia to enhance resistance in local papaya varieties where ring-spot virus is a major pest. Researchers are also modifying other fruits for virus resistance.

Developing Fungi Resistance Food Crops

The search for genetic engineering tactics to combat fungi has intensified with the need to find adequate substitutes for fungicides such as methyl bromide, widely used on fruit and vegetables but being phased out due to its links to ozone depletion.

One emerging area is directed at a plant's production of defensins, a family of naturally occurring antimicrobial proteins which enhance the plant's tolerance to pathogens, especially bacteria.

Certain defensins also demonstrate an ability to fight fungal infections. Defensins are found throughout nature in insects, mammals (including humans), crustaceans, fish and plants.

Defensins from moths and butterflies, the fruit fly, pea seeds and alfalfa seeds all show potent antifungal activity. The first transgenic application of defensins was the incorporation into potatoes of the antifungal defensin from alfalfa. Laboratory and field trials showed that the transgenic potatoes were as resistant to the fungal pathogen Verticillium dahliae as non-transgenic potatoes treated with fungicide.

Although studies are continuing, the chance that fungi will build resistance to defensins is thought unlikely. No known resistant strains of bacteria or fungi have yet evolved that can overcome these highly protective, pesticidal proteins.

On-going research involving banana and cassava is directed to cloning resistance genes for major tropical diseases such as black sigatoka, a leaf fungus that widely infects bananas, cassava mosaic disease and cassava bacterial blight.

In bananas, transgenic lines combining several antifungal genes have been generated. Selected lines are currently being tested for resistance to black sigatoka and Panama disease under greenhouse and field conditions.

Scientists are devising protection against the plant fungus Botrytis cinerea, a serious pathogen in wheat and barley. The strategy uses the gene for a natural plant defence compound named resveratrol. Scientists have also introduced a gene from a wine grape into barley to confer resistance to Botrytis cinerea. Field trials are underway.

Resistance to potato late blight, a disease caused by Phytophthora infestans, receives high priority in potato research. Plant disease from this fungus can be destructive to crop production, as was dramatically illustrated in the Irish potato famine.

In 1995, a U.S. late blight epidemic (caused by new aggressive strains of Phytophthora infestans) affected nearly 160,000 acres of potatoes, or about 20 per cent of domestic production.

Research is underway to genetically engineer potatoes that express the enzyme glucose oxidase and develop resistance to Phytophthora blights (Douches undated). At present, however, no products are close to commercialisation. Potatoes are also being transformed using a soybean gene for a protein (beta-1, 3-endoglucanase) that confers resistance to infection by Phytophthora.

Other studies report that transgenic potatoes expressing a protein called osmotin showed reduced damage from lesion growth in leaves inoculated with the Phytophthora infestans pathogen.

Still other research is attempting to boost fungal resistance in potatoes by transferring resistance genes from peas. Infection of these transgenic potatoes with the fungus triggers hormone-like signals in the potatoes that turn on the pea resistance genes.

One substance that is produced, chitosan, stops fungal growth and activates the potato's own natural defence systems. In rice, blast and sheath blight are major fungal diseases. Scientists created transgenic strains resistant to sheath blight that are currently being field-tested.

Developing Bacteria Resistance Food Crops

Most food crops are susceptible to bacterial diseases, but bacteria rarely attack certain plants, such as mosses, ferns and conifers. Bacterial infections in plants may cause leaf and fruit spots (lesions), soft rots, yellowing, wilting, stunting, tumours, scabs or blossom blights.

When tissue damage occurs on the blossoms, fruit or roots of food crops, yields may be reduced. Potatoes are susceptible to blackleg and soft rot diseases caused by the bacterial pathogen Erwinia carotovora.

To combat these bacteria, scientists have exploited the family of enzymes known as lysozymes that catalyze the breakdown of bacterial cell walls. Using cloned lysozyme genes and a promoter, transgenic potatoes were created that produced lysozyme.

In laboratory tests, the transformed potatoes exhibited substantially enhanced resistance to Erwinia carotovora. Field tests and further development of resistant lines are in progress. A different transgenic strategy to combat Erwinia carotovora was demonstrated in tobacco engineered to overexpress a peptide that kills bacteria.

The genetically engineered tobacco plants were resistant to both Erwinia carotovora and Pseudomonas syringae pv tabaci, the pathogen responsible for wild fire disease in rice. Scientists have also successfully transferred a bacterial resistance gene from wild rice to cultivated rice.

Developing Insects Resistance Food Crops

There are several different combat tactics, including engineering for the expression of toxins in plants that kill insects when they consume the plant material, but are nontoxic to other species that eat the plant. Other alterations focus on inducing sterility in the pest organism or affecting the digestion or metabolism of the pests.

In addition, attempts to enhance a plant's natural ability to produce leaf wax could make the plant more difficult for insects to consume. The best known and most widely used transgenic pest-protected crops are those that express insecticidal proteins derived from genes cloned from the soil bacterium Bacillus thuringiensis, more commonly known as Bt. Crystal (Cry) proteins or delta-endotoxins formed by this bacterium are toxic to many insect species.

Delta-endotoxins bind specifically in the insect gut to receptor proteins, destroying cells and killing the insect in several days (shown below). There are several different Bt strains containing many different toxins. Scientists have identified and isolated the genes for several toxin proteins from different Bt strains.

In recent years, these genes have been introduced into several crop plants in an effort to protect them from insect attack and eliminate the need for spraying synthetic chemical pesticides. There are more than 100 patents for Bt Cry genes. Bt field corn, sweet corn, soy, potato and cotton are commercialized in the U.S., and one or more of these are commercialized in at least 11 other countries.

Bt controls the larvae of butterflies and moths (Lepidopteran insects) that eat the plants. It is especially effective against the larvae of the European corn borer (shown left), a significant corn pest in the U.S., as well as the Southwestern corn borer and the lesser cornstalk borer. In sweet corn, Bt toxins effectively deter corn earworm and fall armyworm.

Recently, a different strain of Bt, Bacillus thuringiensis tenebrionis, was used as a gene source to confer resistance to corn rootworm, another major pest in cornfields. The resistant corn is currently in field trials. Bt hybrid rice is also undergoing field-testing and is showing considerable effectiveness in resisting major pests in Asia such as the leaf folder, yellow stem borer and striped stem borer.

Bt canola is also under development. Borers also create a good environment for fungi to grow. Where fusarium fungi grow, they reduce plant quality and generate fumonisins-toxins that can be fatal to farm animals and have been linked to liver and esophageal cancer in African farmers. Thus, one way to reduce fungal contamination is to control pests.

Scientists have measured reductions in fumonisin levels in Bt corn of 90 percent or greater. Bt works against insects that eat plant tissue. However, those pests that do not eat the leaves, but rather pierce and suck nutrients from the plant, require different defence strategies. These insects include aphids, white flies and stink bugs.

White flies are a major pest in poinsettias, sweet potatoes and cotton. Because these insects do not consume large amounts of plant material, a leading way to combat them is the genetic expression of toxic proteins that are strong enough to kill the pest, yet safe for the plant and non- target organisms.

Avidin in transgenic corn demonstrates a different approach. Avidin is a glycoprotein, an organic compound composed of both a protein and a carbohydrate, and is usually found in egg whites. Avidin is known for chemically tying up the vitamin biotin, making it unavailable as a nutrient. Insects eating transgenic corn modified to produce avidin die from biotin deficiency.

Although this corn was not toxic to mice, further evaluation of its potential for insect toxicity and safety for human consumption is awaited. Plants produce wax as a natural protective coating. Genetic modification can increase the expression of this inherent trait.

Experiments to increase leaf wax are in the early stages, but scientists have already raised wax content by as much as 15-fold. This strategy is aimed at increasing the plant's resistance to both pests and fungal pathogens.

Developing Nematode Resistance Food Crops

The most common of nematode plant parasites found worldwide is the root-knot nematode. Probably every form of plant life, including field crops, ornamentals and trees, is attacked by at least one species of nematode. They are responsible for 10 per cent of global crop losses worth an estimated $80 billion a year.

Transgenic strategies to combat nematodes are emerging. Nematodes are particularly destructive in bananas, soybeans, rice and potatoes. Scientists are fighting these parasitic worms in potato and banana crops using the genes for cystatins, defence proteins that occur naturally in rice and sunflowers. Incorporation of the genes in potatoes produced as much as 70 per cent nematode resistance in field trials.

Nematodes are particularly fond of soybeans. In the U.S., the soybean cyst nematode is considered the most devastating pest. Standard plant breeding led to a highly resistant variety of soybeans from a wild strain, but it did not cross well with modern soybean lines.

Using genetic markers, a means of identifying cells with particular traits, scientists bred plants containing the resistance gene with domesticated varieties, circumventing the poor performance characteristics of the wild variety. While the new varieties are not transgenic, they resulted from combining the use of modern genetic markers with conventional breeding techniques.

Improving Field-crop Production and Soil Management

Improving field-crop production and soil management is another central aim of genetic engineering technology in commodity crops. Applications include crop resistance to herbicides; improved nitrogen utilisation, reducing need for fertiliser; increased tolerance to stresses such as drought and frost; regulation of plant hormones, which are key to plant growth and development; attempts to increase yield, and a multitude of other, less widespread applications.

There are many negative effects when weeds grow with crop plants, the most common being competition for sunlight, water, space and soil nutrients. If weeds grow with crops, they too use these growth factors, and may cause losses great enough to justify control measures.

In addition to economic yield loss, other concerns may determine when weed control is justified. For example, eastern black nightshade in soybeans or late-emerging grasses in corn may not reduce yield, but these weeds can clog equipment, causing harvest delays. The most common method currently employed to manage weeds is the use of herbicides.

The use of genetic modification techniques has created crops that are both tolerant and resistant to herbicides, or weed killers. This technology allows herbicides to be sprayed over resistant crops from emergence through flowering, thus making the applications more effective.

To date, six categories of these crops have been engineered to be resistant to the herbicides glyphosate, glufosinate ammonium, imidazolinone, sulfonylurea, sethoxydim and bromoxynil.

Probably the best-known herbicide for which tolerance has been genetically engineered into crops is glyphosate, known commercially by brand names such as Roundup, Rodeo and Accord. Resistance to glyphosate is the transgenic trait most common in agriculture worldwide. To date soy, corn, cotton, canola, sugar beets and, most recently, wheat, have been genetically transformed for glyphosate tolerance.

Although glyphosate has been used as an herbicide for 26 years, transgenic glyphosate-resistant crops are a more recent development and are widely deployed on acres devoted to soy and cotton. Research is underway to create other glyphosate tolerant crops. To date, two weed species, annual rigid ryegrass and goose grass, have built resistance to glyphosate.

Corn, soy, rice, sugar beet, sweet corn and canola have also been genetically modified to tolerate the herbicide glufosinate ammonium. The seeds for these crops are sold commercially under brand names such as Liberty Link. Transgenic soybeans, cotton and flax with a tolerance to the herbicide sulfonylurea are also on the market.

Other strains of engineered soybeans and corn are resistant to sethoxydim, the active ingredient in the commercial herbicides Poast, Poast Plus, and Headline, used to control undesirable grass species.

The herbicide bromoxynil, sold under the commercial name Buctril, is normally toxic to cotton, a broadleaf crop, and is primarily used on grass-like crops, such as corn, sorghum and small grains, to kill invading broadleaf weeds. Scientists have genetically modified cotton plants for resistance to this herbicide, allowing its use to control broadleaf weeds in cotton fields.

Improved Nitrogen Utilization

There appear to be relatively few biotechnology applications specifically designed to enhance the characteristics of farm crops, such as size, yield, branching, seed size and number. Scientists have, however, created some enhancements. A recent example is the discovery of a gene in the alga Chlorella sorokiniana that has a unique enzyme not found in conventional crop plants.

The enzyme, ammonium-inducible glutamate dehydrogenase, increases the efficiency of ammonium incorporation into proteins. In some plants, it increases the efficiency of nitrogen use. The practical implication is that less fertilizer would be necessary to grow these plants. When the gene was incorporated into wheat, biomass production, growth rate and kernel weight all increased, as did the number of spikes in the plant.

Stress Tolerance

Stress tolerance involves a family of genes, rather than a single one. They are rapidly activated in response to cold, inducing the expression of "cold-regulated" genes, and resulting in enhanced freezing tolerance.

Over-expression of these genes in Arabidopsis-small plants of the mustard family that are commonly used to study plant genetics-increases freezing tolerance and leads to elevated levels of proline and total soluble sugars, substances that protect against cold. Common stress responses in plants involve water retention at the cellular level.

As a result, researchers have given special attention to osmoprotectant molecules, or molecules that hold water, such as sugars, sugar alcohols, certain amino acids (proline) and quaternary amines like glycinebetaine.

Various plants genetically engineered for increased levels of protectant sugar have shown increased drought tolerance. For instance, Arabidopsis and tobacco plants engineered to produce mannitol, a sugar alcohol, withstood high saline conditions and had enhanced germination rates and increased biomass. Other strategies have addressed different stress factors.

Improved cold tolerance and normal germination under high salt was reported in Arabidopsis engineered to express the enzyme choline oxidase. Transgenic rice, engineered to express the late embryogenesis abundant protein gene transferred from barley, was significantly more tolerant to drought and salinity than conventional varieties of rice.

Another transgenic rice engineered in the laboratory for enhanced expression of the enzyme glutamine synthetase had increased photorespiration capacity and increased tolerance to salt. Preliminary results suggested enhanced tolerance to chilling as well.

Regulation of Plant Hormones

Plant hormones such as auxin, cytokinins, gibberellins, abscisic acid, ethylene, etc. have been targeted for genetic modification to influence plant growth and development- fruit development and ripening; stem elongation and leaf development; germination, dormancy and tolerance of adverse conditions.

These hormone classes are highly interactive; the concentration of one affects the activity of another. For example, the ratio of the hormone abscisic acid to gibberellin in a plant determines whether a seed will remain dormant or germinate.

Recent discovery of an enzyme involved in the production of the hormone auxin enabled researchers to investigate the effects of moderating auxin production in determining plant characteristics. When auxin is overproduced, branching is inhibited and leaves curl down as the plant elongates, a reaction typically related to reduced light exposure. The same gene that produces this enzyme is apparently related to a gene in mammals that governs enzymes that detoxify certain chemicals.

In wheat, the hormone abscisic acid slows seed germination and improves the tolerance to cold and drought. Extending or enhancing the production of abscisic acid may also delay germination, a useful characteristic in climates where spring rain is sparse or falls late in the season.

Production of abscisic acid is increased in response to environmental stress, and a family of enzymes called protein kinases stimulates its production. Selecting plant varieties high in abscisic acid, or engineering plants to produce more of the hormone, may confer greater drought and cold tolerance.

Introduction of dwarfed, high-yielding wheat contributed to the 'Green Revolution' of the 1960s and 1970s, during which world wheat yields almost doubled. Shorter varieties of wheat grains, with a greater resistance to damage by wind, resulted from a reduced response to the hormone gibberellin.

Scientists have since shown that the gene called Rht can cause "dwarfing" in a range of plants, opening up the possibility of quickly developing higher-yielding varieties in several crops. Researchers believe that this strategy could be applied to a still wider range of crops through genetic engineering.

The plant hormone ethylene regulates ripening in fruits and vegetables. Controlling the amount and timing of ethylene production can initiate or delay ripening, which might reduce spoilage that can occur between the time produce is picked and brought to market.

Transgenic techniques aim to regulate the enzyme that breaks down a precursor of ethylene production. By regulating the timing and rate of this degradation, ripening can be controlled. This technology has been applied and field- tested in tomatoes, raspberries, melons, strawberries, cauliflower and broccoli, but has not yet been commercialized.

Breeding for Drought Stress Tolerance

Breeding for drought resistance is the process of breeding plants with the goal of reducing the impact of dehydration on plant growth.

Dehydration Stress in Crop Plants

In nature or crop fields, water is often the most limiting factor for plant growth. If plants do not receive adequate rainfall or irrigation, the resulting dehydration stress can reduce growth more than all other environmental stresses combined.

Drought can be defined as the absence of rainfall or irrigation for a period of time sufficient to deplete soil moisture and cause dehydration in plant tissues. Dehydration stress results when water loss from the plant exceeds the ability of the plant's roots to absorb water and when the plant's water content is reduced enough to interfere with normal plant processes.

Dehydration Stress is a Global Phenomenon

About 15 million km2 of the land surface is covered by crop-land, and about 16% of this area is equipped for irrigation. Thus, in many parts of the world, including the United States, plants may frequently encounter dehydration stress. Rainfall is very seasonal and periodic drought occurs regularly. The effect of drought is more prominent in sandy soils with low water holding capacity. On such soils some plants may experience dehydration stress after only a few days without water.

During the 20th century, the rate of increase in `blue' water withdrawal (from rivers, lakes, and aquifers) for irrigation and other purposes was higher than the growth rate of the world population. Country-wise maps of irrigated areas are available.

Dehydration Stress and Future Challenges to Crop Production

Soil moisture deficit is a significant challenge to the future of crop production. Severe drought in parts of the U.S., Australia, and Africa in recent years drastically reduced crop yields and disrupted regional economies. Even in average years, however, many agricultural regions, including the U.S. Great Plains, suffer from chronic soil moisture deficits. Cereal crops typically attain only about 25% of their potential yield due to the effects of environmental stress, with dehydration stress the most important cause. Two major trends will likely increase the frequency and severity of soil moisture deficits.

Climate Change

Higher temperatures are likely to increase crop water use due to increased transpiration. A warmer atmosphere will also speed up melting of mountain snow pack, resulting in less water available for irrigation. More extreme weather patterns will increase the frequency of drought in some regions.

Competing uses for Limited Water Supplies

Increased demand from municipal and industrial users will further reduce the amount of water available for irrigated crops.

Although changes in tillage and irrigation practices can improve production by conserving water, enhancing the genetic tolerance of crops to drought stress is considered an essential strategy for addressing moisture deficits.

Dehydration Stress Affects Plant Physiology

A plant responds to a lack of water by halting growth and reducing photosynthesis and other plant processes in order to reduce water use. As water loss progresses, leaves of some species may appear to change colour — usually to blue-green. Foliage begins to wilt and, if the plant is not irrigated, leaves will fall off and the plant will eventually die. Soil moisture deficit lowers the water potential of a plant's root and, upon extended exposure, abscisic acid is accumulated and eventually stomatal closure occurs. This reduces a plant's leaf relative water content. The time required for dehydration stress to occur depends on the water-holding capacity of the soil, environmental conditions, stage of plant growth, and plant species. Plants growing in sandy soils with low water-holding capacity are more susceptible to dehydration stress than plants growing in clay soils. A limited root system will accelerate the rate at which dehydration stress develops. A plant's root system may be limited by the presence of competing root systems from neighbouring plants, by site conditions such as compacted soils or high water tables, or by container size (if growing in a container). A plant with a large mass of leaves in relation to the root system is prone to drought stress because the leaves may lose water faster than the roots can supply it. Newly planted plants and poorly established plants may be especially susceptible to dehydration stress because of the limited root system or the large mass of stems and leaves in comparison to roots.

Dehydration Stress Interaction with Other Stress Factors

Aside from the moisture content of the soil, environmental conditions of high light intensity, high temperature, low relative humidity and high wind speed will significantly increase plant water loss. The prior environment of a plant also can influence the development of dehydration stress. A plant that has been exposed to dehydration stress (hardened) previously and has recovered may become more drought resistant. Also, a plant that was well-watered prior to being water-limited will usually survive a period of drought better than a continuously dehydration-stressed plant.

Mechanisms of Drought Resistance

The degree of resistance to drought depends upon individual crops. Generally three strategies can help a crop to mitigate the effect of dehydration stress:

Avoidance

If the plant shows dehydration avoidance, the environmental factor is excluded from the plant tissues by reducing water loss ("water savers", e.g. by thick leaf epicuticular wax, leaf rolling, leaf posture) or maintaining water uptake ("water spenders", e.g. by growing deeper roots). Dehydration avoidance is desirable in modern agriculture, where drought resistance requires the maintenance of economically viable plant production under dehydration stress. The role of dehydration avoidance is maintaining water supply and sustaining leaf hydration and turgidity with the purpose of maintaining stomatal opening and transpiration as long as possible under water deficit. This is essential for leaf gas exchange, photosynthesis and plant production through carbon assimilation.

Tolerance

If the plant shows dehydration tolerance, the environmental factor enters the plant tissues but the tissues survive, by e.g. maintaining turgor and osmotic adjustment.

Escape

Dehydration escape involves e.g. early maturing or seed dormancy, where the plant uses previous optimal conditions to develop vigor.

Dehydration Recovery refers to some plant species being able to recuperate after brief drought periods.

A proper timing of life-cycle, resulting in the completion of the most sensitive developmental stages while water is abundant, is considered to be a dehydration escape strategy. Avoiding dehydration stress with a root system capable of extracting water from deep soil layers, or by reducing evapotranspiration without affecting yields, is considered as dehydration avoidance. Mechanisms such as osmotic adjustment (OA) whereby a plant maintains cell turgor pressure under reduced soil water potential are categorised as dehydration tolerance mechanisms. Dehydration avoidance mechanisms can be expressed even in the absence of stress and are then considered constitutive. Dehydration tolerance mechanisms are the result of a response triggered by dehydration stress itself and are therefore considered adaptive. When the stress is terminal and predictable, dehydration escape through the use of shorter duration varieties is often the preferable method of improving yield potential. Dehydration avoidance and tolerance mechanisms are required in situations where the timing of drought is mostly unpredictable.

Drought resistance mechanisms are genetically controlled and genes or QTL responsible for drought resistance have been discovered in several crops which opens avenue for molecular breeding for drought resistance.

Drought Resistance Traits

Resistance to drought is a quantitative trait, with a complex phenotype, often confounded by plant phenology. Breeding for drought resistance is further complicated since several types of abiotic stress, such as high temperatures, high irradiance, and nutrient toxicities or deficiencies can challenge crop plants simultaneously.

Osmotic Adjustment

When a plant is exposed to water deficit, it may accumulate a variety of osmotically active compounds such as amino acids and sugars, resulting in a lowering of the osmotic potential. Examples of amino acids that may be up-regulated are proline and glycine betaine. This is termed osmotic adjustment and enables the plant to take up water, maintain turgor and survive longer.

Cell Membrane Stability

The ability to survive dehydration is influenced by a cell's ability to survive at reduced water content. This can be considered complementary to OA because both traits will help maintain leaf growth (or prevent leaf death) during water deficit. Crop varieties differ in dehydration tolerance and an important factor for such differences is the capacity of the cell membrane to prevent electrolyte leakage at decreasing water content, or "cell membrane stability (CMS)". The maintenance of membrane function is assumed to mean that cell activity is also maintained. Measurements of CMS have been used in different crops and are known to be correlated with yields under high temperature and possibly under dehydration stress.

Epicuticular Wax

In sorghum (Sorghum bicolor L. Moench), drought resistance is a trait that is highly correlated with the thickness of the epicuticular wax layer. Experiments have demonstrated that rice varieties with a thick cuticle layer retain their leaf turgor for longer periods of time after the onset of a water-stress.

Partitioning and Stem Reserve Mobilisation

As photosynthesis becomes inhibited by dehydration, the grain filling process becomes increasingly reliant on stem reserve utilisation. Numerous studies have reported that stem reserve mobilisation capacity is related to yield under dehydration stress in wheat. In rice, a few studies also indicated that this mechanism maintains grain yield under dehydration stress at the grain filling stage. This dehydration tolerance mechanism is stimulated by a decrease in gibberellic acid concentration and an increase in abscisic acid concentration.

Manipulation and Stability of Flowering Processes

Seedling Drought Resistance Traits

For emergence from deep sowing (to exploit dry upper soil), this is practised to help seedlings reach the receding moisture profile, and to avoid high soil surface temperatures which inhibit germination. Screening at these stage provides practical advantages, specially when managing large amount of germ-plasms.

The Drought Resistant Ideotype

Usually ideotypes are developed to create an ideal plant variety. The following traits constitutes ideotype of wheat by CIMMYT:

1. Large seed size: Helps emergence, early ground cover, and initial biomass.

2. Long coleoptiles: For emergence from deep sowing.

3. Early ground cover.

Thinner, wider leaves (i.e., with a relatively low specific leaf weight) and a more prostrate growth habit help to increase ground cover, thus conserving soil moisture and potentially increasing radiation use efficiency.

1. High pre-anthesis biomass.

2. Good capacity for stem reserves and remobilisation.

3. High spike photosynthetic capacity.

4. High RLWC/Gs/CTD during grain filling to indicate ability to extract water.

5. Osmotic adjustment.

6. Accumulation of ABA.

The benefit of ABA accumulation under dehydration has been demonstrated. It appears to pre-adapt plants to stress by reducing stomatal conductance, rates of cell division, organ size, and increasing development rate. However, high ABA can also result in sterility problems since high ABA levels may abort developing florets.

1. Heat Tolerance: The contribution of heat tolerance to performance under dehydration stress needs to be quantified, but it is relatively easy to screen for.

2. Leaf anatomy: Waxiness, pubescence, rolling, thickness, posture. These traits decrease radiation load to the leaf surface. Benefits include a lower evapotranspiration rate and reduced risk of irreversible photo-inhibition. However, they may also be associated with reduce radiation use efficiency, which would reduce yield under more favourable conditions.

3. High tiller survival: Comparison of old and new varieties have shown that under dehydration older varieties over-produce tillers many of which fail to set grain while modern drought resistant lines produce fewer tillers most of which survive.

4. Stay-green: The trait may indicate the presence of drought resistance mechanisms, but probably does not contribute to yield per se if there is no water left in the soil profile by the end of the cycle to support leaf gas exchange. It may be detrimental if it indicates lack of ability to remobilise stem reserves. However, research in sorghum has indicated that Stay-green is associated with higher leaf chlorophyll content at all stages of development and both were associated with improved yield and transpiration efficiency under dehydration.

Combination Phenomics: Overall Health of Crops

The concept of combination phenomics comes from the idea that two or more plant stresses have common physiological effects or common traits - which are an indicator of overall plant health. As both biotic and abiotic stresses can result in similar physiological consequence, drought resistant plants can be separated from sensitive plants. Some imaging or infrared measuring techniques can help to speed the process for breeding process. For example, spot blotch intensity and canopy temperature depression can be monitored with canopy temperature depression.

Molecular Breeding for Drought Resistance

Recent research breakthroughs in biotechnology have revived interest in targeted drought resistance breeding and use of new genomics tools to enhance crop water productivity. Marker-assisted breeding is revolutionising the improvement of temperate field crops and will have similar impacts on breeding of tropical crops. Other molecular breeding tool include development of genetically modified crops that can tolerate plant stress. As a complement to the recent rapid progress in genomics, a better understanding of physiological mechanisms of dehydration response will also contribute to the progress of genetic enhancement of crop drought resistance. It is now well accepted that the complexity of the dehydration syndrome can only be tackled with a holistic approach that integrates physiological dissection of crop dehydration avoidance and - tolerance traits using molecular genetic tools such as MAS, micro-arrays and transgenic crops, with agronomic practices that lead to better conservation and utilisation of soil moisture, and better matching of crop genotypes with the environment. MAS has been implemented in rice varieties to assess the drought tolerance and to develop new abiotic stress-tolerant varieties.

Breeding for Heat Stress Tolerance

Plant breeding is process of development of new cultivars. Plant breeding involves development of varieties for different environmental conditions – some of them are not favorable. Among them, heat stress is one of such factor that reduces the production and quality significantly. So breeding against heat is a very important criterion for breeding for current as well as future environments produced by global climate change (e.g. global warming).

Breeding for Heat Stress Tolerance in Plants

Heat stress due to increased temperature is a very important problem globally.[citation needed] Occasional or prolonged high temperatures cause different morpho-anatomical, physiological and biochemical changes in plants. The ultimate effect is on plant growth as well as development and reduced yield and quality. Breeding for heat stress tolerance can be mitigated by breeding plant varieties that have improved levels of thermo-tolerance using different conventional or advanced genetic tools. Marker assisted selection techniques for breeding are highly useful. Recently 41 polymorphic SSR markers has been identified between a heat tolerant rice variety 'N22' and heat susceptible-high yielding variety 'Uma' for the development of new 'high yielding-heat tolerant' rice varieties.

What is Heat Stress Tolerance

Heat stress is defined as increased temperature level sufficient to cause irreversible damage to plant growth and development. Generally a temperature rise, above usually 10 to 15 °C above ambient, can be considered heat shock or heat stress. Heat tolerance is broadly defined as the ability of the plant tolerate heat – means that grow and produce economic yield under high temperatures.

Significance: Current and Future - Global Warming

Heat stress is a serious threat to crop production globally. Global warming is particularly consequence of increased level of green house gases such as CO_2, methane, chlorofluorocarbons and nitrous oxides. The Intergovernmental Panel on Climatic Change (IPCC) has predicted a rise of 0.3 °C per decade reaching to approximately 1 and 3 °C above the present value by 2025 and 2100 AD, respectively.

Physiological Consequence of Heat Stress

At very high temperatures cause severe cellular injury and cell death may occur within short time, thus leading to a catastrophic collapse of cellular organization. However, under moderately high temperatures, the injury can only occur after longer exposure to such a temperature however the plant efficiency can be severely affected. High temperature directly affect injuries such as protein denaturation and aggregation, and increased fluidity of membrane lipids. Other indirect or slower heat injuries involve inactivation of enzymes in chloroplast and mitochondria, protein degradation, inhibition of protein synthesis, and loss of membrane integrity. Heat stress associated injuries ultimately lead to starvation, inhibition of growth, reduced ion flux, production of toxic compounds and production of reactive oxygen species (ROS). Immediately after exposure to high temperature stress-related proteins are expressed as stress defense strategy of the cell. Expression of heat shock proteins (HSPs), protein with 10 to 200 kDa, is supposed to be involved in signal transduction during heat stress. In many species it has been demonstrated that HSPs results in improved physiological phenomena such as photosynthesis, assimilate partitioning, water and nutrient use efficiency, and membrane stability.

Studies have found tremendous variation within and between species, thus this will help to breed heat tolerance for future environment. Some of attempts to develop heat-tolerant genotypes are successful.

Traits Associated with Heat Stress Tolerance

Different physiological mechanisms may contribute to heat tolerance in the field—for example, heat tolerant metabolism as indicated by higher photosynthetic rates, stay-green, and membrane thermo-stability, or heat avoidance as indicated by canopy temperature depression. Several physiological and morphological traits have been evaluated for heat tolerance - Canopy temperature, leaf chlorophyll, stay green, leaf conductance, spike number, biomass, and flowering date.

a) Canopy Temperature Depression (CTD): CTD has shown clear association with yield in warm environments shows it association with heat stress tolerance. CTD shows high genetic correlation with yield and high values of proportion of direct response to selection indicating that the trait is heritable and therefore amenable to early generation selection. Since an integrated CTD value can be measured almost instantaneously on scores of plants in a small breeding plot (thus reducing error normally associated with traits measured on individual plants), work has been conducted to evaluate its potential as an indirect selection criterion for genetic gains in yield. CTD is affected by many physiological factors, which makes it a powerful.

b) Stomatal Conductance: Canopy temperature depression is highly suitable for selecting physiologically superior lines in warm, low relative humidity environments where high evaporative demand leads to leaf cooling of up to 10 °C below ambient temperatures. This permits differences among genotypes to be detected relatively easily using infrared thermometry. However, such differences cannot be detected in high relative humidity environments because the effect of evaporative cooling of leaves is negligible. Nonetheless, leaves maintain their stomata open to permit the uptake of CO_2, and differences in the rate of CO_2 fixation may lead to differences in leaf conductance that can be measured using a porometer. Porometry can be used to screen individual plants. The heritability of stomatal conductance is reasonably high, with reported values typically in the range of 0.5 to 0.8. Plants can be assessed for leaf conductance using a viscous flow porometer that is available on the market (Thermoline and CSIRO, Australia). This instrument can give a relative measure of stomatal conductance in a few seconds, making it possible to identify physiologically superior genotypes from within bulks.

c) Membrane Thermostability: Although resistance to high temperatures involves several complex tolerance and avoidance mechanisms, the membrane is thought to be a site of primary physiological injury by heat, and measurement of solute leakage from tissue can be used to estimate damage to membranes. Since membrane thermostability is reasonably heritable and shows high genetic correlation with yield.

d) Chlorophyll Fluorescence: Chlorophyll fluorescence, an indication of the fate of excitation energy in the photosynthetic apparatus, has been used indicator for heat stress tolerance.

e) Chlorophyll Content and Stay Green: Chlorophyll content and stay green traits have been found to be associated with heat stress tolerance.,. Xu et al. identified three QTLs for chlorophyll content (Chl1, Chl2, and Chl3) (coincided with three stay-green QTL regions (Stg1, Stg2, and Stg3)) were identified in Sorghum. The Stg1 and Stg2 regions also contain the genes for key photosynthetic enzymes, heat shock proteins, and an abscisic acid (ABA) responsive gene.

f) Photosynthesis: Decliend photosynthesis is suggested as measure of heat stress sensitivity in plants.

g) Stem Reserve Remobilization.

Combination Breeding and Physiological Breeding

The physiological-trait-based breeding approach has merit over breeding for yield per se because it increases the probability of crosses resulting in additive gene action. The concept of combination phenomics comes from the idea that two or more stress have common physiological effect or common traits - which is an indicator of overall plant health. Similar analogy in human medical terms is high blood pressure or high body temperature or high white blood cells in body is an indicator of health problems and thus we can select healthy people from unhealthy using such a measure. As both abiotic and abiotic stresses can result in similar physiological consequence, tolerant plant can be separated from sensitive plants. Some imaging or infrared measuring techniques can help to speed the process for breeding process. For example, spot blotch intensity and canopy temperature depression can be monitored with canopy temperature depression.

Molecular Farming in Plants

Plant molecular farming describes the production of recombinant proteins and other secondary metabolites in plants. This technology depends on a genetic transformation of plants that can be accomplished by the methods of stable gene transfer, such as gene transfer to nuclei and chloroplasts, and unstable transfer methods like viral vectors. An increasing quest for biomedicines has coincided with the high costs and inefficient production systems (bacterial, microbial eukaryotes, mammalian cells, insect cells, and transgenic animals). Therefore, transgenic plants as the bioreactors of a new generation have been the subject of considerable attention with respect to their advantages, such as the safety of recombinant proteins (antibodies, enzymes, vaccines, growth factors, etc.), and their potential for the large-scale and low-cost production. However, the application of transgenic plants can entail some worrying concerns, namely the amplification and diffusion of transgene, accumulation of recombinant protein toxicity in the environment, contamination of food chain, and costs of subsequent processing. The given threats need to be the subject of further caution and investigation to generate valuable products, such as enzymes, pharmaceutical proteins, and biomedicines by the safest, cheapest, and most efficient methods.

Molecular farming is a biotechnological program that includes the genetic modification of agricultural products to produce proteins and chemicals for commercial and pharmaceutical purposes. A vast majority of developing countries cannot afford the high costs of medical treatments resulted from the existing methods. Hence, we need to produce not only the new drugs but also the cheaper versions of the present samples in the market. Molecular farming can offer efficient solutions for the current growing need for the biomedicines. Plants provide an inexpensive and simple system for the production of valuable recombinant proteins on large scale, and compared to the other production systems, they have numerous advantages in terms of economy, safety, and applicability. Though using transgenic plants has entailed some sorts of limitations and concerns, the optimization has been operated for solving the existing problems. Normally, the production of pharmaceutical proteins has been largely concentrated by the technology of molecular farming in plants, also plants can be used for the production of food supplements, biopolymers, industrial enzymes, and proteins in the investigations (avidin, β-glucuronidase, etc.). Prior production systems, including bacteria, microbial eukaryotes (yeasts, double-stranded fungi), animal cells, and transgenic animals, as a result of their limitations, were replaced by transgenic plants.

The Strategies of Plant Transformation

Plant molecular farming depending on the production of transgenic plants has been operated by two general methods as the following:

Stable or Permanent Expression Systems

a. Stable nuclear transformation: Stable nuclear transformation refers to the integration of genes or nominated foreign genes into the nuclear genome of plants, which results in the change of genetic structures and consequent expression of transgenes after integration with the host genomes. The largest amount of recombinant proteins has been produced by one of the most common methods of stable nuclear transformation. A method exploited for aggregating proteins in dried beans of maize culminates in a long-term storage in the beans at the room temperature without decomposition of proteins. In addition, it has a considerable potential for producing crops like cereals that actually grow everywhere. However, a long production cycle for some crops and their potential collisions with natural species or food products have restricted the wide acceptance of this method.

b. Stable plastid transformation: Plastid transformation offers a remarkable solution in comparison to that of nuclear transformation since it has numerous advantages including preventing transgene escape through amphimixis (because plastids are inherited through the maternal tissue in the majority of species.) and absence of chloroplasts in pollen and consequent improbability of their transfer, which reduces environmental concerns. The transformed transgenic plants with homoplasmic chloroplasts (all chloroplasts carry transgenes) were selected after several generations of plant regeneration from bombarded leaf explants. Selection was conducted on a medium containing spectinomycin or combined with streptomycin. The researchers have already extracted a human pharmaceutical protein, more than 3% to 6%, from the total soluble proteins in the chloroplasts of tobacco. Recently, Oey et al. reported a very high level (70% of an entire soluble protein) for a protein antibiotic with the chloroplast system, which, till today, has been the highest concentration of recombinant proteins. Despite this, the great potential of plastid transformation has some functional limitations. Although this technology has been developed in other species such as tomatoes, lettuce, soy, and eggs , in the current situation, chloroplast transformation only in tobacco is practically possible, but unfortunately this plant is inedible and full of poisonous alkaloids; in addition, long lasting storage in refrigerators will bring about changes in protein stability.

c. Plant cell suspension culture: This method involves the removal of cell walls and gene transfer to the obtained protoplasts and suspension culture. The purification system and its downstream processing are cheaper and easier. In addition, the use of suspension culture can decrease heterogeneity in proteins and sugar (N-glycans) regarding the uniformity of the type and size of cells. Furthermore, as a fast system there is no need for the production of transgenic plants; however, the cell lines can be produced after a few months.

Temporary or Transient Expression Systems

A transient production may be the fastest system for plant molecular farming. Nowadays, these are the systems routinely applied for verifying expression constructs during a few weeks for a significant amount of proteins. The given systems include the following methods:

Agrobacterium transformation method: Infiltration of recombinant agrobacterium suspension into tobacco leaf tissue is achieved without stable gene transfer, which facilitates the transfer of T-DNA to a very high percentage of cells, where the transgenes are expressed at a high level without a stable transfer of genes. Presently, this method has been very efficient for the production of clinical biomedicines with a fast expansion.

Viral infection methods: The viral infection method depends on the capability of plant viruses, such as tobacco mosaic virus and X potato virus, which functions as a vector to convey foreign genes into plant genomes without fusing with the genome of that plant.

The Limitations and Optimization of Plant Production Systems

Optimization of Expression of Transcripts

To optimize the expression of transcripts, a widely used strategy is the use of building promoters, such as cauliflower mosaic virus 35S RNA promoter and maize 1-ubiquitin promoter, respectively, suitable for spilt-cotyledons and single-cotyledons. Tissue-specific and organ-specific promoters are used for stimulating the expression of transgenes (antigen vaccine HBsAgM, single-chain variable fragment Maureen G4, and Human interferon-α) in some tissues or organs, such as tubers, seeds, and fruits. The given specific expression of tissues prevents the accumulation of recombinant proteins in vegetative organs, which can have a negative impact on plant growth; for example, palatine is a gland-specific promoter; i.e., the protein is expressed in the gland but not in leaves; and also ubiquitin promoter is specified for the embryonic tissues of plants. Transcription factors (e.g., AlcR) can act as the invigorator of promoters to increase the level of transgene expression. The stability of transcripts of genes can be achieved by co-expression of the specified gene and an RNA silencing inhibitor.

Optimization of Translation

Expression constructs can be designed for guaranteeing the efficiency of translation and the sustainability of transcripts. As an instance, the removal of 5' untranslated region and natural ´3 for foreign genes and introducing the leader sequence of tobacco mosaic virus RNA, RUB13 rice polyubiquitin gene, alfalfa mosaic virus, or tobacco viruses in the expressions, all, individually, have shown a significant increase in the level of transgene expressions.

In addition to the leader sequences, expression cassettes can be designed with the AU-rich sequences in 3' untranslated regions, which may change or be removed as the editing sites for ensuring the stability of transcript. It is also proved that every organism shows codon usage deviations that may be the subject of importance for adapting the coding sequence of heterologous genes for the host gene to optimize the efficiency of translation. In this regard, the site of initial translation from heterologous protein to pair with Kozak consensus sequence, with the application of GCTTCCTCC sequence, started after codon or ACC, or ACA had been changed before that. It is better to unscientifically estimate codon changes rather than their real amount considering the changes in the expression level of transgenes in similar systems and the use of similar structures. To this end, an increase of codon combinations of (A/G)(a/c)(a/g)AUG and (A/g)(u/C)(g/C)AUG for the optimal operation of translation was, respectively, reported in Arabidopsis and rice. The given change in transgene expression could be due to the position effect, number of transgene copies, or gene silencing.

Regarding the effect of position, expression cassettes can be designed to have nuclear matrix attachment regions for ensuring the transgene insertion in proper sites for stimulating transcription factors for promoters. Furthermore, the problem of position effect can be prohibited by targeting the transgene to plastids. To optimize the production of single-cotyledon transgene, the strategies that include the use of specific genetic elements containing cAMP response elements for a simultaneous transfer with transgene in T-DNA are used. In addition, one new technology, including the structure of an artificial autonomous mini-chromosome, can genetically materialize excellent possibilities with several advantages, namely genetic stability due to the absence of gene silencing and position effect.

Optimization of Protein Stability

To optimize the stability of recombinant proteins, known as the most important limiting factor for the function of molecular farming, the targeting of proteins into certain intracellular parts is demanded. The intracellular targeting not only increases protein stability but also determines the processing type of dependent protein. This can be applied for the optimization of isolation and procedures of purification by the fusion proteins and targets with high affinity. Targeting of proteins can be done by the following pathways and organelles:

- The intracellular parts, like protein storage vacuoles, have been discovered for the accumulation of recombinant proteins.

- Cathepsin D inhibitor can act as the agent of stability of protein structures to protect the targeted recombinant proteins in the cytosol of plants. Recombinant protein production through this signal has been proved to be very effective and economical.

- To protect proteins from cytosolic degradation, these proteins can be targeted by fusion to a C-terminal tail without a forced passing through the lumen of the endoplasmic reticulum to the membrane surface. To enhance the ease of purification, proteins can be fused to oleosin proteins as oil bodies in order to target protein expression with the oil bodies of seeds.

- The proteins, like in glycosylation, that do not need post-translation modifications for their activity, can be targeted to chloroplast since post-translation modifications are not conducted in these organelles.

- Targeting for accumulation in endoplasmic reticulum is accomplished by two methods: one is adding SEKDEL endoplasmic reticulum signals to the end of C-protein, and the other is using fused N or C signals with y-zein. Endoplasmic reticulum is an oxidizing environment with high amounts of chaperone proteins and low levels of proteases. This pathway is suitable for the proteins that need post-translational modifications (e.g., glycosylation). The breakdown of proteins by proteases (proteolytic degradation) outside the cell is another noticeable factor for investigating the plant-based production of biomedicines.

Challenge of Glycosylation (Protein Quality)

Glycosylation refers to the covalent binding of sugars to proteins in order to increase close-packing, biological activity, solubility, and biological functionality. Glycosylation takes place in plants in the secretory pathway of endoplasmic reticulum and golgi apparatus. The glycosylation patterns of plants and animals differ in the composition of N-glycans; plants add residues of α (1, 3) fucose and β-(1,2) xylose to N-glycans of their protein, but animals add residues of (1 and 6) fucose, glucose, and sialic acid to N-glycans. These differences can be problematic for humans when medical animal proteins extracted from plants are used; consequently, a correct human N-glycosylation demands a plant engineering. A number of strategies for changing the pattern of N-glycosylation in plants have been elaborated as following:

- The use of purified enzymes of β-(1,4) galactosyltransferase and Sialyltransferase for making glass transitions in the recombinant proteins derived from plants.

- Co-expression of β-(1,4) galactosyltransferase human enzyme with the target transgene in transgenic plants.

- Prohibiting the activity of fucosyltransferase and xylyltransferase enzymes.

- Targeting pharmaceutical proteins to the endoplasmic reticulum in order to avoid the addition of protective N-glycans.

Selecting Appropriate Host Plants

Major economical factors in appointing an appropriate host include the total biomass yield, storage characteristics, ease of transport, value of recombinant proteins, maintenance costs, its availability for workers, required area, duration of production cycle, cost of subsequent products, and edibility. In addition to the economical analysis, a sufficient host should be appropriate in terms of transformation and regeneration. In addition to the high potential of tobacco for transformation and regeneration, it has the majority of the aforementioned economic benefits. However, tobacco (except the cultivar 81 V9) contains high amounts of toxic combinations, nicotine and other alkaloids, that cannot be removed during the purification process. In spite of this, alternative forage crops like alfalfa and lettuce are being investigated and discovered as a suitable host for molecular farming. However, forage plants generally suffer from the problem of instability of expressed proteins, by which drying and freezing of the leaves and immediate processing following the harvest have been inevitable. The seed-based expression of proteins is considered to be more ideal regarding the fact that it neither affects the growth of plants nor needs the freezing of leaves or immediate processing after harvest, and it allows the long-term storage of proteins at a limited temperature without decreasing the level of activity. In this regard, grains, especially rice and corn, have been cited as the superior ones. Maize has abundant advantages, such as having the highest rate of biomass yield among food crops and ease of transformation and production increase. The high amount of protein (20%-40%) in the grains of legumes with remarkable levels of self-pollination in soy and peas is the main reason for transgenes of these plants for protein accumulation.

Predicting the Intracellular Localization of the Recombinant Protein

The importance of intracellular localization of proteins is due to the functional consequences of proteins. Therefore, the problem of intracellular localization of amino acid sequences has been the subject of great attention in the community of bioinformatics. Thus, various methods, like searching for targeted signals, have been presented with respect to a prediction that various proteins are produced in different intercellular segments.

Proteins and Biomedicines Produced in Plants

Plants are able to produce those bacterial and viral recombinant antigens that preserve the capability of making the structures Type IV similar to those witnessed in mammalian systems, and the post-translational modifications are operated to maintain the biological activity of proteins. The most important issue is vaccine production in the edible tissues of transgenic plants, which is a very safe and effective method in vaccination.

The biomedicines produced in plants are as follows:

- Antigens for the production of edible vaccines: Antigens, used for generating an immune response resulting in immunity against diseases in human proteins, are expressed from different pathogens in plants. Those vaccines derived from plants have been so far induced immunity against rabies virus, hepatitis B, rotavirus, HIV, and other pathogens.

- Monoclonal antibodies: Widespread application of antibodies has lead to the study of new methods in order to strengthen efficiency and reduce the cost of producing antibodies.

Among the studied methods, using transgenic plants as bioreactors are known as the most efficient one. While designing therapeutic antibodies in the production of recombinant expression systems, the apprehension of the functioning mechanisms of antibodies is essential. Although the primary function of antibodies is actualized by binding to antigens, it does not act as a protective performance. Some antibodies have a direct neutralizing impact, for instance blocking the bacteria or the active sites of the pathogenic factors such as enzymes. The antibodies produced in plants incorporate Immunoglobulin G (IgG) and Immunoglobulin A (IgA), IgA and IgG shimmer molecules, IgG and IgA secreted molecules, Single-Chain variable fragment, fragment antigen-binding, and second variable of heavy and light chains.

- Pharmaceutical proteins: Some samples of biomedicines recently expressed in plants include erythropoietin, interferon, hirudin, aprotinin, Leu-enkephalin, somatotropin of human growth hormone.

- Non-pharmaceutical proteins derived from plants or industrial proteins belong mainly to the enzymes that include avidin, trypsin, aprotinin, β-glucocerebrosidase, peroxidase and cellulose, etc., listed by Basaran and Rodriguez-Cerezo and now available in the market. Molecular farming of destructive enzymes of the cell walls such as cellulose, hemicellulase, xylanase, and particularly ligninase provide a great status for the biofuel industry respecting cellulosic ethano.

Molecular Farming and Metabolic Engineering

An Opportunity for Producing Plants with a High Technology

Molecular farming and metabolic engineering make the production of new high-tech products possible. There is a driving force backing molecular farming that makes its costs much less than traditional farming. Chlamydomonas reinhardtii, as a unicellular alga, is one of the most recent production projects examined by Franklin and Mayfield. C. Reinhardtii is the only plant whose transformation was operated in its all segments containing DNA (nucleus, plastid, and mitochondria). Unique features of the moss system bring about the possibility of removing target genes and purification of the proteins secreted from the culture medium. The target gene was omitted to get rid of the nuclear genes for glycosylation. The first step towards the long-term goals of reengineering mechanism in modifications of plant proteins is setting a new standard in all systems of plant expressions in order to humanize the biomedicines produced in plants.

Purification and Downstream Processing of the Recombinant Proteins

Recovery usually includes the process and breakdown of plant tissues, protein extraction, solid-liquid separation, and protein concentration while purification encompasses safety protection, liquid-liquid extraction, membrane filtration, chromatography, etc. The processing of leaves requires a particular attention; leaves should be processed immediately after the harvest or frozen to prevent protein degradation by proteases, whereas seeds can be stored for a long period of time due to the less probability of destruction of recombinant proteins expressed in seeds. Using the secretory systems of cells can also be beneficial since disintegrating plant cells throughout recovery is not required; thus, the release of phenolic compounds can be avoided while the recombinant proteins can be unstable in culture mediums. Another way of facilitating the recovery of proteins is utilizing continuous labels. Protein labels must be removed after purification so that the structure of purified protein can change into its natural position. The technology of oleosin fusion, through which the gene sequence of recombinant proteins is fused to the sequence of a special internal oil protein called oleosin in safflower and canola, is separated after the digestion of internal protease following protein purification.

Costs of Subsequent Processing

The costs of subsequent processing of the recombinant proteins derived from plants have been estimated about 80% of the total production costs. This is why so much attention has been paid at sufficient strategies for reducing the costs to the least amount. The application of watery textures like tomatoes as a production system has been expanded because of their potential for reducing the costs via the ease of extracting from their textures in comparison with those of dry tissues like grains. In addition, tomatoes are highly regarded as a reputed host crop in terms of its bio-safety because these plants grow in greenhouses without worrying about the preservation of transgenic plants.

Nowadays, oil bodies of oilseed agricultural products, like the seeds of safflower and mustard, are being exploited by the application of oleosin fusion technology developed by SemBioSys Genetics in order to facilitate the purification of recombinant proteins and reduction of subsequent costs. The strategies including targeting of recombinant proteins for the seeds of oilseed agricultural products as an oleosin fusion facilitate the extraction of fused proteins from oil bodies and the release of the recombinant proteins from their fusion partner; one example can be the accumulation and purification of biologically active human insulin, apolipoprotein A-I (Milano) and human growth hormone in safflower.

There are several recombinant proteins derived from plants that were the basic idea of edible vaccines, directly eaten as fruits (tomatoes and bananas) and vegetables (lettuce and carrots); accordingly, no processing costs will be demanded by the elimination of processing,. Bananas, as a fruit host in agricultural products, have particularly attracted lots of customers for the production of edible vaccines, especially for developing countries. This has been widely developed in such countries because of long distance transports and cooling requirements. Apart from the mentioned advantages, high digestibility and palatability of bananas have won a wide public acceptance for the vaccination of children. The sufficiency of potatoes, eaten in raw or low processed forms, for edible vaccines has resulted in their wide production. Potatoes, like seeds, have the advantage of production stability due to a special molecular environment allocated in glands.

Bio-safety and the Challenges in the Domain of Protein Production

Biomedicines in Molecular Farming

The risks of transgenic plants are divided into two categories: one category directly affects humans and the other endangers environment and other organisms. The attack of immune system can disable these medicines and lead to the stimuli for the allergic reactions, some of which have been elaborated as follows:

- There are some concerns in terms of environmental pollution about the entrance of transgenes into the food chain, which requires a sound management and supervision.

- The other concern refers to the grain transformations using agrobacterium since grains are important crops in the production of pharmaceutical protein.

- The reactions of immune system can disable the medicines produced in plants and be the stimuli for allergic reactions.

References

- Plant-breeding-steps-and-methods-of-plant-breeding-for-disease-resistance, plants: biologydiscussion.com, Retrieved 19 July, 2019

- Plant-breeding, science: britannica.com, Retrieved 17 August, 2019

- Siebert, S; Döll, P; Hoogeveen, J; Faures, J M; Frenken, K; Feick, S (2005). "Development and validation of the global map of irrigation areas". Hydrol. Earth Syst. Sci. 9 (5): 535–47. doi:10.5194/hess-9-535-2005

- Transgenic-plants-meaning-reasons-and-fundamentals, transgenic-plants: biotechnologynotes.com, Retrieved 2 January, 2019

- Salam, S. A., Sindhumole, P., Waghmare, S. G., & Sajini, S. (2017). Molecular characterization of rice (Oryza sativa L.) genotypes for drought tolerance using two SSR markers. Electronic Journal of Plant Breeding, 8(2), 474-479

- Applications-of-transgenic-plants-6-applications, transgenic-plants, plants: biologydiscussion.com, Retrieved 6 February, 2019

- Disease-resistance-in-plants-5-techniques-biotechnology, genetically-modified, biotechnology: biotechnologynotes.com, Retrieved 9 March, 2019

- Rosyara, U. R., S. Subdedi, R. C. Sharma and E. Duveiller. 2010. Photochemical Efficiency and SPAD Value as Indirect Selection Criteria for Combined Selection of Spot Blotch and Terminal Heat Stress in Wheat. Journal of Phytopathology Volume 158, Issue 11-12, pages 813–821, December 2010

- Molecular-farming-in-plants, plants-for-the-future: intechopen.com, Retrieved 27 April, 2019

Chapter 4
Molecular Marker-aided Breeding

A fragment of DNA which is associated with a certain location within a genome is known as a molecular marker. They are used in biotechnology and molecular biology in order to make the breeding process much more efficient and speedy. This chapter has been carefully written to provide an easy understanding of the various applications of molecular markers for the purpose of breeding.

Molecular Marker

A molecular marker is a DNA sequence in the genome which can be located and identified. As a result of genetic alterations (mutations, insertions, deletions), the base composition at a particular location of the genome may be different in different plants.

These differences, collectively called as polymorphisms can be mapped and identified. Plant breeders always prefer to detect the gene as the molecular marker, although this is not always possible. The alternative is to have markers which are closely associated with genes and inherited together.

The molecular markers are highly reliable and advantageous in plant breeding programmes:

1. Molecular markers provide a true representations of the genetic makeup at the DNA level.

2. They are consistent and not affected by environmental factors.

3. Molecular markers can be detected much before development of plants occur.

4. A large number of markers can be generated as per the needs.

Basic Principle of Molecular Marker Detection

Let us assume that there are two plants of the same species—one with disease sensitivity and the other with disease resistance. If there is DNA marker that can identify these two alleles, then the genome can be extracted, digested by restriction enzymes, and separated by gel electrophoresis. The DNA fragments can be detected by their separation. For instance, the disease resistant plant may have a shorter DNA fragment while the disease — sensitive plant may have a longer DNA fragment.

Molecular markers are of two types:

1. Based on nucleic acid (DNA) hybridization (non-PCR based approaches).

2. Based on PCR amplification (PCR-based approaches).

Markers based on DNA Hybridization

The DNA piece can be cloned, and allowed to hybridize with the genomic DNA which can be detected. Marker-based DNA hybridization is widely used. The major limitation of this approach is that it requires large quantities of DNA and the use of radioactivity (labeled probes).

Restriction Fragment Length Polymorphism (RFLP)

RFLP was the very first technology employed for the detection of polymorphism, based on the DNA sequence differences. RFLP is mainly based on the altered restriction enzyme sites, as a result of mutations and re-combinations of genomic DNA. An outline of the RFLP analysis is given in figure. The procedure basically involves the isolation of genomic DNA, its digestion by restriction enzymes, separation by electrophoresis, and finally hybridization by incubating with cloned and labeled probes.

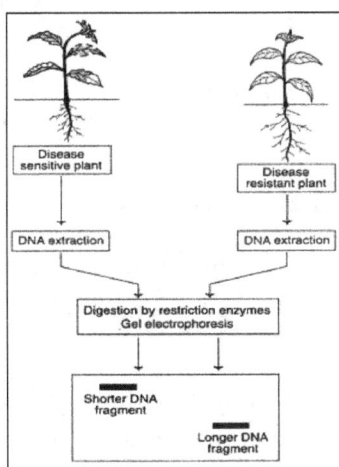

Basic principle of molecular marker detection (screening of genotypes for the identification of DNA markers).

Based on the presence of restriction sites, DNA fragments of different lengths can be generated by using different restriction enzymes. In the figure two DNA molecules from two plants (A and B) are shown. In plant A, a mutations has occurred leading to the loss of restriction site that can be digested by EcoRI.

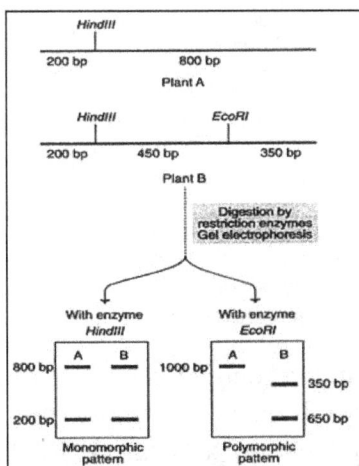

A schematic representation of restriction fragment length polymorphism (RFPL) analysis as molecular marker.

The result is that when the DNA molecules are digested by the enzyme Hindlll, there is no difference in the DNA fragments separated. However, with the enzyme EcoRI, plant A DNA molecules is not digested while plant B DNA molecule is digested. This results in a polymorphic pattern of separation.

Markers based on PCR Amplification

Polymerase chain reaction (PCR) is a novel technique for the amplification of selected regions of DNA. The advantage with PCR is that even a minute quantity of DNA can be amplified. Thus, PCR-based molecular markers require only a small quantity of DNA to start with.

PCR-based markers may be divided into two types:

1. Locus non-specific markers e.g. random amplified polymorphic DNA (RAPD); amplified fragment length polymorphism (AFLP).

2. Locus specific markers e.g. simple sequence repeats (SSR); single nucleotide polymorphism (SNP).

Random Amplified Polymorphic DNA (RAPD) Markers

RAPD is a molecular marker based on PCR amplification. An outline of RAPD is depicted in figure. The DNA isolated from the genome is denatured the template molecules are annealed with primers, and amplified by PCR.

Single short oligonucleotide primers (usually a 10-base primer) can be arbitrarily selected and used for the amplification DNA segments of the genome (which may be in distributed throughout the genome). The amplified products are separated on electrophoresis and identified.

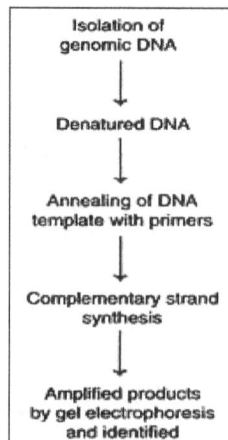

An outline of random amplified poly-morphic DNA (RAPD)
analysis as a molecular marker in plant breeding.

Based on the nucleotide alterations in the genome, the polymorphisms of amplified DNA sequences differ which can be identified as bends on gel electrophoresis. Genomic DNA from two different plants often results in different amplification patterns i.e. RAPDs. This is based on the fact that a particular fragment of DNA may be generated from one individual, and not from others. This represents polymorphism and can be used as a molecular marker of a particular species.

Amplified Fragment Length Polymorphism (AFLP)

AFLP is a novel technique involving a combination of RFLP and RAPD. AFLP is based on the principle of generation of DNA fragments using restriction enzymes and oligonucleotide adaptors (or linkers), and their amplification by PCR. Thus, this technique combines the usefulness of restriction digestion and PCR.

The DNA of the genome is extracted. It is subjected to restriction digestion by two enzymes (a rare cutter e.g. MseI; a frequent cutter e.g. EcoRI). The cut ends on both sides are then ligated to known sequences of oligonucleotides.

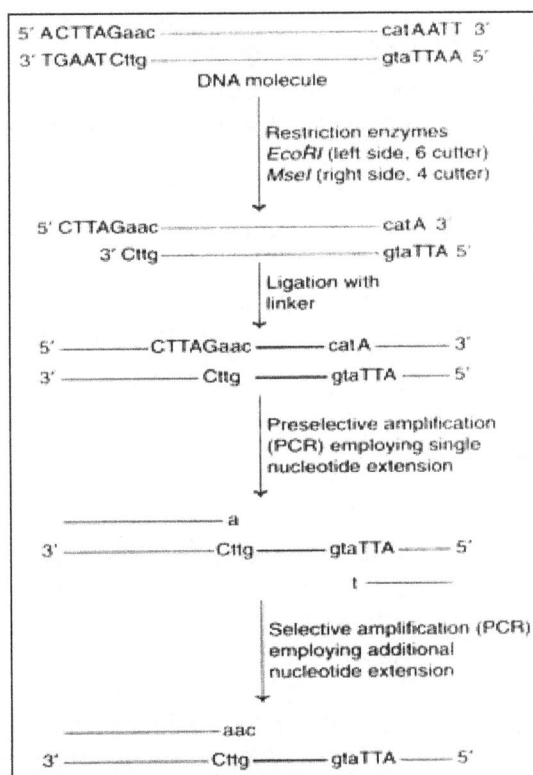

A diagrammatic representation of the amplified fragment length polymorphism (AFPL).

PCR is now performed for the pre-selection of a fragment of DNA which has a single specific nucleotide. By this approach of pre-selective amplification, the pool of fragments can be reduced from the original mixture. In the second round of amplification by PCR, three nucleotide sequences are amplified.

This further reduces the pool of DNA fragments to a manageable level (< 100). Autoradiography can be performed for the detection of DNA fragments. Use of radiolabeled primers and fluorescently labeled fragments quickens AFLP.

AFLP analysis is tedious and requires the involvement of skilled technical personnel. Hence some people are not in favour of this technique. In recent years, commercial kits are made available for AFLP analysis. AFLP is very sensitive and reproducible. It does not require prior knowledge of sequence information. By AFLP, a large number of polymorphic bands can be produced and detected.

Sequence Tagged Sites (STS)

Sequence tagged sites represent unique simple copy segments of genomes, whose DNA sequences are known, and which can be amplified by using PCR. STS markers are based on the polymorphism of simple nucleotide repeats e.g. $(GA)_n$, $(GT)_n$, $(CAA)_n$ etc. on the genome. STS have been recently developed in plants. When the STS loci contain simple sequence length polymorphisms (SSLPs), they are highly valuable as molecular markers. STS loci have been analysed and studied in a number of plant species.

Microsatellites

Microsatellites are the tandemly repeated multi-copies of mono-, di-, tri- and tetra nucleotide motifs. In some instances, the flanking sequence of the repeat sequences may be unique. Primers can be designed for such flanking sequences to detect the sequence tagged microsatellites (STMS). This can be done by PCR.

Sequence Characterized Amplified Regions (SCARs)

SCARs are the modified forms of STS markers. They are developed by PCR primer that are made for the ends of RAPD fragment. The STS-converted RAPD markers are sometimes referred to as SCARs. SCARs are useful for the rapid development of STS markers.

Molecular Marker Assisted Selection

Selection of the desired traits and improvement of crops has been a part of the conventional breeding programmes. This is predominantly based on the identification of phenotypes. It is now an accepted fact that the phenotypes do not necessarily represent the genotypes. Many a times the environment may mark the genotype. Thus, the plant's genetic potential is not truly reflected in the phenotypic expression for various reasons.

The molecular marker assisted selection is based on the identification of DNA markers that link/represent the plant traits. These traits include resistance to pathogens and insects, tolerance to abiotic stresses, and various other qualitative and quantitative traits. The advantage with a molecular marker is that a plant breeder can select a suitable marker for the desired trait which can be detected well in advance. Accordingly, breeding programmes can be planned.

The following are the major requirements for the molecular marked selection in plant breeding:

1. The marker should be closely linked with the desired trait.

2. The marker screening methods must be efficient, reproducible and easy to carry out.

3. The analysis should be economical.

Molecular Breeding

With rapid progress in molecular biology and genetic engineering, there is now a possibility of improving the crop plants with respect to yield and quality. The term molecular breeding is frequently used to represent the breeding methods that are coupled with genetic engineering techniques.

Improved agriculture to meet the food demands of the world is a high priority area. For several years, the conventional plant breeding programmes (although time consuming) have certainly helped to improve grain yield and cereal production.

The development of dwarf and semi-dwarf varieties of rice and wheat have been responsible for the 'Green Revolution', which has helped to feed millions of poverty-stricken people around the world. Many developments on the agriculture front are expected in the coming years as a result of molecular breeding.

Linkage Analysis

Linkage analysis basically deals with studies to correlate the link between the molecular marker and a desired trait. This is an important aspect of molecular breeding programmes. Linkage analysis has to be carried out among the populations of several generations to establish the appropriate linkage. In the earlier years, linkage analysis was carried out by use of isoenzymes and the associated polymorphisms. Molecular markers are now being used. The techniques employed for this purpose have already been described.

Quantitative Trait Loci

These are many characteristics controlled by several genes in a complex manner. Some good examples are growth habit, yield, adaptability to environment, and disease resistance. These are referred to as quantitative traits. The locations on the chromosomes for these genes are regarded as quantitative trait loci (QTL).

The major problem, the plant breeder faces is how to improve the a complex character controlled by many genes. It is not an easy job to manipulate multiple genes in genetic engineering. Therefore, it is a very difficult and time consuming process. For instance, development of Golden Rice (with enriched pro-vitamin A) involving the insertion of just three genes took about seven years.

Arid and Semi-arid Plant Biotechnology

The terms arid zone is used to refer to harsh environmental conditions with extreme heat and cold. The fields have limited water and minerals. It is different task to grow plants and achieve good crop yield in arid zones. Semi-arid regions are characterized by unpredictable weather, inconsistent rainfall, long dry seasons, and poor nutrients in the soil.

The biotechnological approaches for the breeding programmes in the semi-arid regions should cover the following areas:

1. Development of crops that are tolerant to drought and salinity.

2. Improvements to withstand various biotic and abiotic stresses.

3. Micro-propagation techniques to spread economically important plants which can withstand harsh environmental conditions.

Some success has been achieved in improving sorghum, millet and legume crops that are grown in semi-arid regions. Genetic transformation in sorghum was possible by using micro projectile method.

Greenhouse and Green-home Technology

Greenhouse literally means a building made up of glass to grow plants. Green houses are required to grow regenerated plants for further propagation and for growing plants to maturity. Greenhouses are the intermediary stages involving the transitional step between the plant cultures and plant fields. The purpose of greenhouses is to acclimatize and test the plants before they are released into the natural environment.

The plants are grown in greenhouse to develop adequate root systems and leaves so as to withstand the field environment. The greenhouses are normally equipped with cooling systems to control temperature. Greenhouses have chambers fitted with artificial lights. It is possible to subject the plants to different lighting profiles. In recent years many improvements have been made in the development of more suitable greenhouses. These include the parameters such as soil, and humidity.

The major limitation of greenhouse technology is an increase in CO_2 production that in turn increases temperature. Some approaches are available to control temperature. Green home technology is a recent development. In this case, temperature is controlled by using minimum energy.

Molecular Markers and Molecular Breeding in Plants

The advent of molecular techniques played a significant role in increase our knowledge of cereal genetics and behaviour of cereal genomics. While RFLP markers have been the basis for most work in crop plants, valuable markers have been generated from RAPD and AFLPs.

Recently, other improvised molecular marker such as simple sequence repeats (SSR), microsatellite marker have also been developed for major crop plants and initiate rapid advance in both marker development and implementation in breeding programme.

Conventional plant breeding is time consuming and depends on environmental conditions, for example, breeding for new variety development takes between eight and twelve years without any guarantee of variety release.

Hence, breeding is extremely enthusiastic and extremely interested in new molecular breeding technologies that could make this entire breeding exercise simple, speedy and efficient. This technology also offers selection of desirable combination of traits.

This approach can be establishing linkage between molecular marker and traits to be selected. Once this approach is completed would allow breeding process to be conducted in the laboratory without waiting for expression of the genes for particular phenotype. For example, resistant to plant pathogen can be evaluated in the absence of disease. Every stress tolerance can be analysed in the seedling stages itself.

Marker Assisted Selection (MAS)

1.   Mapping of plant genomes:

     Among several important crop plant, rice has been the target plant for intense mapping studies. However, model plant, Arabidopsis thaliana has also been considered for extensive mapping study. Currently, genetic mapping for rice and Arabidopsis has also been completed.

Mapping smaller the size of genome is ideal for key strategy for the identification of gene in the complex genome. Molecular marker like RFLP marker from closely related species is good reliable markers for constructing gene map. Microsatellite is useful tool for constructing gene map to each species whereas RFLP loci can be used to map at greater degree in related taxa.

2. Linkage of molecular marker to desired trait:

   Identification of genes responsible for useful trait may be established by a linkage analysis with markers on a genetic map of plant genome polymorphic marker is generally used to identify linked markers. Finding of linked markers can be accomplished by a certain useful technique like bulked segregate analysis (BSA). This technique is used to detect polymorphism between two DNA samples made up of bulk of individuals from the segregating populations.

   For example, one bulk sample DNA from one individual contain target gene while other DNA from individual devoid of this gene. The segregating population derived sample contains both of the two bulks containing most genes.

   Polymorphism between the bulks is likely to be linked to genes for the trait. The analysis of linkers for dominant (RAPD) and co-dominant marker (RFLP or microsatellite) requires separate unique analysis of mapping with F_2 population.

3. Accelerated back crossing:

   Marker assisted selection facilitate in the acceleration of whole breeding process, allowing earlier release of plant commercial. This is achieved by two important method like accelerated back crossing and selection for a desired trait. Introduction of desirable trait has been the favourite choice for plant breeders without altering other character.

This can be accomplished by repeated crossing to the plant with the genetic background required. Each and every generation requires selection of introduced trait. This requires number of crossing and more number of generation since molecular marker facilitate selection of individuals with more of the recurrent genome at each generation. It can limit breeding programme to be completed with few generation.

Selection for a Desired Trait

Several desirable traits can be directly selected by molecular marker and can be screen at any stage in the breeding programme. In addition to the available marker for routine use, conversion of RFLP marker to PCR based marker helps significantly in the economy of molecular usage. In addition, usage of molecular marker for fruit selection in plant breeding requires availability of simple, inexpensive technique that provides rapid result in the assessment for next round of selection.

Molecular Breeding for Resistance

PCR-based marker has been used in the breeding of desirable resistance to viral and fungal pathogen in plants. Barley yellow mosaic virus (Ba YMU) has been considered as important viral disease in Europe. Therefore, breeding for resistance to the disease is special importance. Applications of closely linked PCR-based markers for the transmission of resistance gene(s) against barley yellow mosaic virus are now successful and efficient.

Similarly breeding for resistance to fungal pathogen has been advanced. Fusarium head blight is a serious disease of wheat. Molecular markers closely liked to the major QTL involved in Fusarium head blight (FHB) resistance have been identified and raise the possibility of marker assisted selection (MAS) for introducing resistance alleles into elite wheat variety. These are some of the safety strategy in breeding. The new varieties combined high yield performance and high level of resistance to fusarium pathogen.

Identification of Breeding Lines

In the germplasm labelling error can lead to breeding artifact because handling of large number of lines may create problem in identifying molecular marker can be used to confirm breeding lines.

Identification of Hybridity

Molecular marker can be used to identify hybrid nature of individual especially in self- pollinating species. Production of hybrid through non-conventional hybrid method like somatic hybrid also can be identical using RAPD analysis.

Purity of Breeding Lines

Accidental mixing of seeds or cross-contamination in seed harvested may lead to contamination of breeding lines. Molecular markers can be used to assist establishment of pure breeding lines and check contamination of breeding.

Prediction of Hybrid Performance (Heterosis)

Establishment of genetic distance between the parents used in the cross can able to ascertain the performance of hybridity. Genetic distance between possible parents can be estimated by employing molecular markers. RFLP microsatellite markers are selected as useful marker for these predictions.

Identifying Germplasm

Identification of several useful genetic resources of possible parents for use in breeding requires suitable molecular technique. RAPD marker is useful tool for the survey of germplasm, for example, survey of rice germplasm using RAPD shows linkage between the presence of specific marker and quantitative trait loci (QTL) for novel character.

Biochemical Markers

Before the onset of molecular markers, some of the earlier experiments were carried out using biochemical markers. Certain isozymes (or isoenzymes) have been employed as biochemical markers in various aspects of plant breeding and genetics due to their significance as natural markers.

Some of the commonly known biochemical isozyme markers are esterases, peroxidases, dehydrogenases etc. Basically these markers are gene expression products and are characterised by electrophoresis and staining. By definition, isoenzymes are multiple molecular forms of the same enzyme that execute the same function.

They are the products of the different alleles of one or several genes. In several cases monomer and dimer isozymes are most often employed due to their early segregation process. Biochemical marker and assessment showed that these are co-dominant markers. Although isoenzymes are potentially a reliable marker, their polymorphism is however, exhibit relatively poor within a cultivated species.

Random Amplified Polymorphic DNA Markers and its Applications

The random amplified polymorphic DNA (RAPD) technique based on the polymerase chain reaction (PCR) has been one of the most commonly used molecular techniques to develop DNA markers. RAPD is a modification of the PCR in which a single, short and arbitrary oligonucleotide primer, able to anneal and prime at multiple locations throughout the genome, can produce a spectrum of amplification products that are characteristics of the template DNA. RAPD markers have found a wide range of applications in gene mapping, population genetics, molecular evolutionary genetics, and plant and animal breeding. This is mainly due to the speed, cost and efficiency of the technique to generate large numbers of markers in a short period compared with previous methods.

Advances in molecular biology techniques have provided the basis for uncovering virtually unlimited numbers of DNA markers. Over the last decade, polymerase chain reaction (PCR) has become a widespread technique for several novel genetic assays based on selective amplification of DNA. This popularity of PCR is primarily due to its apparent simplicity and high probability of success. Unfortunately, because of the need for DNA sequence information, PCR assays are limited in their application. The discovery that PCR with random primers can be used to amplify a set of randomly distributed loci in any genome facilitated the development of genetic markers for a variety of purposes. RAPD (pronounced 'rapid'), for Random Amplification of Polymorphic DNA, is a type of PCR reaction, but the segments of DNA that are amplified are random. The RAPD analysis described by Williams et al. is a commonly used molecular marker in genetic diversity studies. No knowledge of the DNA sequence for the targeted gene is required, as the primers will bind somewhere in the se quence, but it is not certain exactly where. This makes the method popular for comparing the DNA of biological systems that have not had the attention of the scientific community, or in a system in which relatively few DNA sequences are compared (it is not suitable for forming a DNA databank). Perhaps the main reason for the success is the gain of a large number of genetic markers that require small amounts of DNA without the requirement for cloning, sequencing or any other form of the molecular characterisation of the genome of the species in question.

RAPD markers are decamer (10 nucleotide length) DNA fragments from PCR amplification of random segments of genomic DNA with single primer of arbitrary nucleotide sequence and which are able to differentiate between genetically distinct individuals, although not necessarily in a reproducible way. It is used to analyze the genetic diversity of an individual by using random primers. In this paper, the principles, working mechanism, differences between standard PCR and RAPD -PCR, characteristics, laboratory steps, data analysis and interpretation, advantages and disadvantages and several of the most common applications of RAPD markers in biology are discussed.

Principle of the RAPD Technique

The principle is that, a single, short oligonucleotide primer, which binds to many different loci, is used to amplify random sequences from a complex DNA template. This means that the amplified fragment generated by PCR depends on the length and size of both the primer and the target genome. The assumption is made that a given DNA sequence (complementary to that of the primer) will occur in the genome, on opposite DNA strands, in opposite orientation within a distance that is readily amplifiable by PCR. These amplified products (of up to 3.0 kb) are usually separated on agarose gels (1.5-2.0%) and visualised by ethidium bromide staining.

The use of a single decamer oligonucleotide promotes the generation of several discrete DNA products and these are considered to originate from different genetic loci. Polymorphisms result from mutations or rearrangements either at or between the primer binding sites and are detected as the presence or absence of a particular RAPD band. This means that RAPDs are dominant markers and, therefore, cannot be used to identify heterozygotes.

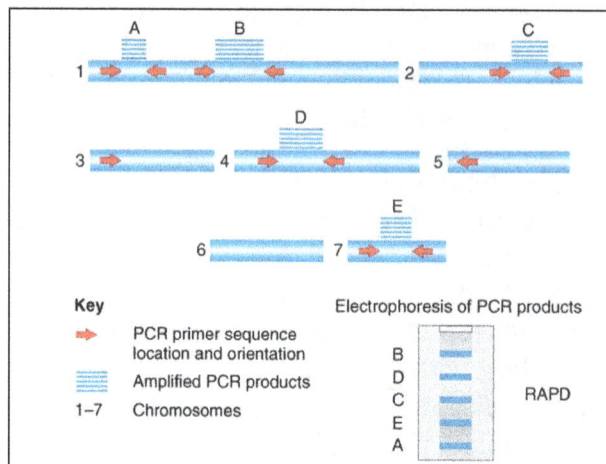

RAPD analysis general model.

The standard RAPD utilises short synthetic oligonucleotides (10 bases long) of random sequences as primers to amplify nanogram amounts of total genomic DNA under low annealing temperatures by PCR. Primers are commercially available from various sources (e.g. Operon Technologies Inc., California; Biosciences, Bangalore; Eurofinns, Bangalore; GCC Biotech, Kolkata).

PCR amplification with primers shorter than 10 nucleotides [DNA amplification finger printing (DAF)] has also been used to produce more complex DNA fingerprinting profiles. Although these approaches are different with respect to the length of the random primers, amplification conditions and visualization methods, they all differ from the standard PCR condition in that only a single oligonucleotide of random sequence is employed and no prior knowledge of the genome subjected to analysis is required.

At an appropriate annealing temperature during the thermal cycle, oligonucleotide primers of random sequence bind several priming sites on the complementary sequences in the template genomic DNA and produce discrete DNA products if these priming sites are within an amplifiable distance of each other. The profile of amplified DNA primarily depends on nucleotide sequence homology between the template DNA and oligonucleotide primer at the end of each amplified product. Nucleotide variation between different sets of template DNAs will result in the presence or absence of bands because of changes in the priming sites. Recently, sequence characterised amplified regions (SCARs) analysis of RAPD polymorphisms showed that one cause of RAPD polymorphisms is chromosomal rearrangements such as insertions/deletions. Therefore, amplification products from the same alleles in a heterozygote differ in length and will be detected as presence and absence of bands in the RAPD profile, which is similar to that of low stringency minisatellite DNA fingerprinting patterns and is therefore also termed RAPD fingerprinting. On average, each primer directs amplification of several discrete loci in the genome so that allelism is not distinguishable in RAPD patterns.

Differences between Standard PCR and RAPD PCR

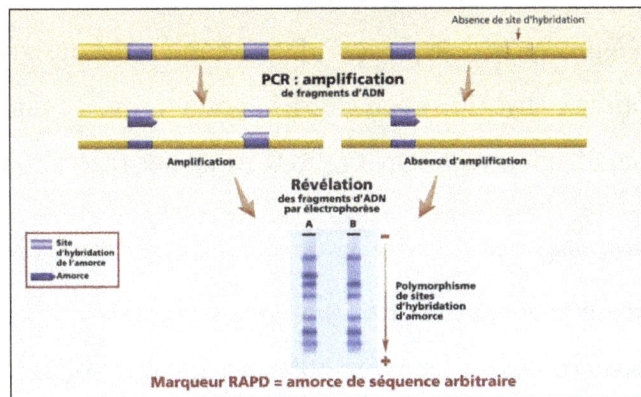

RAPD PCR

In RAPD analysis, the target sequence(s) (to be amplified) is unknown. A primer is designed with an arbitrary sequence. In order for PCR to occur: 1) the primers must anneal in a particular orientation (such that they point towards each other) and, 2) they must anneal within a reasonable distance of one another. Figure depicts a RAPD reaction, a large fragment of DNA (genome A) is used as the template in a PCR reaction containing many copies of a single arbitrary primer.

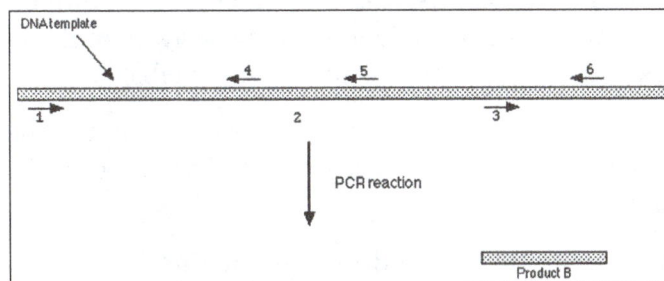

RAPD reaction for genome A. The arrows represent multiple copies of a primer (all primers have the same sequence). The direction of the arrow also indicates the direction in which DNA synthesis will occur. The numbers represent locations on the DNA template to which the primers anneal. Primers anneal to sites 1, 2, and 3 on the bottom strand of the DNA template and primers anneal to sites 4, 5, and 6 on the top strand of the DNA template.

RAPD reaction 1 for genome A. In the above example , only 2 RAPD PCR products are formed:

- Product A is produced by PCR amplification of the DNA sequence which lies in between the primers bound at positions 2 and 5.

- Product B is produced by PCR amplification of the DNA sequence which lies in between the primers bound at positions 3 and 6. Note that no PCR product is produced by the primers bound at positions 1 and 4 because these primers are too far apart to allow completion of the PCR reaction. Also no PCR products are produced at positions 4 and 2 or positions 5 and 3 because these primer pairs are not oriented towards each other.

Advantages of RAPD

RAPD has been used widely because of the following advantages:

- It requires no DNA probes and sequence information for the design of specific primers.

- It involves no blotting or hybridisation steps, hence, it is quick, simple and efficient.

- It requires only small amounts of DNA (about 10 ng per reaction) and the procedure can be automated.

- High number of fragments.

- Arbitrary primers are easily purchased.

- Unit costs per assay are low compared to other marker technologies.

Disadvantages of RAPD

- Nearly all RAPD markers are dominant, i.e. it is not possible to distinguish whether a DNA segment is amplified from a locus that is heterozygous (1 copy) or homozygous (2 copies). Codominant RAPD markers, observed as different-sized DNA segments amplified from the same locus, are detected only rarely.

- PCR is an enzymatic reaction, therefore, the quality and concentration of template DNA, concentrations of PCR components, and the PCR cycling conditions may greatly influence the outcome. Thus, the RAPD technique is notoriously laboratory-dependent and needs carefully developed laboratory protocols to be reproducible.

- Mismatches between the primer and the template may result in the total absence of PCR product as well as in a merely decreased amount of the product. Thus, the RAPD results can be difficult to interpret.

- Lack of a prior knowledge on the identity of the amplification products.

- Problems with reproducibility (sensitive to changes in the quality of DNA, PCR components and PCR conditions).

- Problems of co-migration (do equal-sized bands correspond to the same homologous DNA fragment?). Gel electrophoresis can separate DNA quantitatively, cannot separate equal-sized fragments qualitatively (i.e. according to base sequence).

Developing Locus-specific, Codominant Markers from RAPDS

- The polymorphic RAPD marker band is isolated from the gel.

- It is amplified in the PCR reaction. The PCR product is cloned and sequenced.

- New longer and specific primers are designed for the DNA sequence, which is called the sequenced characterized amplified region marker (SCAR).

Applications of RAPD Analysis

It has become widely used in the study of:

- Genetic diversity/polymorphism,

- Germplasm characterization,

- Genetic structure of populations,

- Domestication,

- Detection of somaclonal variation,

- Cultivar identification,

- Hybrid purity,

- Genome mapping,

- Developing genetic markers linked to a

- Trait in question,

- Population and evolutionary genetics,

- Plant and animal breeding,

- Animal-plant-microbe interactions,

- Pesticide/herbicide resistance,

- Animal behavior study,

- Forensic studies.

RAPD markers exhibit reasonable speed, cost and efficiency compared with other methods; and RAPD can be done in a moderate laboratory. Therefore, despite its reproducibility problem, it will probably be important until better techniques are developed in terms of cost, time and labour.

Chapter 5
Plant Transformation Technology

The method used to insert DNA from another plant, into the genome of a plant of interest is known as plant transformation. Plant transformation vectors are the plasmids that are created to facilitate the generation of transgenic plants. The chapter closely examines these key concepts of plant transformation technology to provide an extensive understanding of the subject.

Plant Transformation

Plant transformation is a way to insert DNA from another organism- normally another plant, into the genome of a plant of interested. For example, a gene called Stilbene synthase is Intsert from grape into tomato to enable the production of the medicinal compound resveratrol in high levels in tomatoes.

Genetic modification is used in many areas of scientific research and is widely used to modify: Yeast, bacteria, plants and mammalian cells. One of the most important uses of genetically modified organisms is for the large scale production of medicinally important compounds. For example, genetically modified bacteria is used for the production of synthetic insulin.

Within plants we use plant transformation to study the effect of certain genes, and to improve plant traits such as: yield, disease resistance, stress tolerance, and nutrient production. Engineering plants for the future, by improving stress tolerance (to stresses such as high salt levels, and drought) is an especially important area of research- with the foreseeable problems farmers are soon to encounter with the changing climate due to global warming.

There are two types of plant transformation : Stable transformation, and transient transformation. This post will focus on stable transformation.

What is Stable Plant Transformation?

Stable plant transformation is used for the stable introduction of a gene into a plant- meaning the gene will be fully integrated in the host genome, so it is expressed continuously, and will also be expressed in later generations of the plant. This type of plant transformation is used for the longer-term research of genes, and for long-term production of a trait/compound in large scale.

Methods used for Stable Transformation

Plant transformation involves 2 stages: Delivery of the DNA into a single cell, and regeneration into full fertile plants.

DNA Delivery

There are 2 main methods for gene delivery in plants- Agrobacterium and the particle gun.

Agrobacterium

Agrobacterium is a naturally occurring soil bacteria, which has the unique ability to transfer part of its own DNA into plant cells. In the wild, transfer of some of the bacterial DNA causes rapid plant cell division, and the development of a plant tumour.

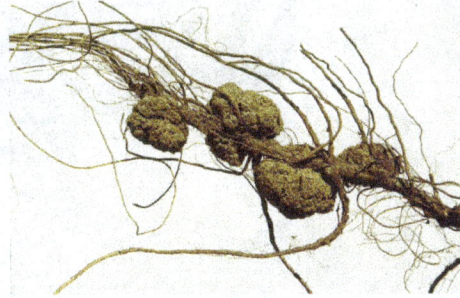

Agrobacterium Gall (tumor) at the root of Carya illinoensis.

Plant Transformation using Particle Bombardment

The Particle bombardment device, also known as the gene gun, was developed to enable penetration of the cell wall so that genetic material containing a gene of interest can be transferred into the cell. Today the gene gun is used for genetic transformation of many organisms to introduce a diverse range of desirable traits.

Plant transformation using particle bombardment follows the same outline as Agrobacterium-mediated method. The steps taken include: 1) isolate the genes of interest from the source organism; 2) develop a functional transgenic construct including the gene of interest; promoters to drive expression; codon modification, if needed to increase successful protein production; and marker genes to facilitate tracking of the introduced genes in the host plant; 3) incorporate into a useful plasmid; 4) introduce the transgenes into plant cells; 5) regenerate the plants cells; and 6) test trait performance or gene expression at lab, greenhouse and field level.

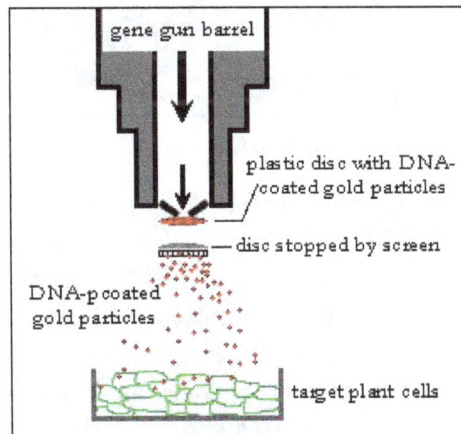

Diagrammatic illustration of gene transfer using Gene Gun method.

The particle bombardment method starts with coating tungsten or gold particles (microprojectiles) with plasmid DNA. The coated particles are coated on a macro-projectile, which is accelerated with air pressure and shot into plant tissue on a petri plate. A perforated plate is used to stop the macro-projectile, while allowing the microprojectiles to pass through to the cells on the other side. As the microprojectiles enter the cells, the transgenes are released from the particle surface and may incorporate into the chromosomal DNA of the cells. Selectable markers are used to identify the cells that take up the transgene. The transformed plant cells are then regenerated into whole plants using tissue culture.

Helios Gene Gun.

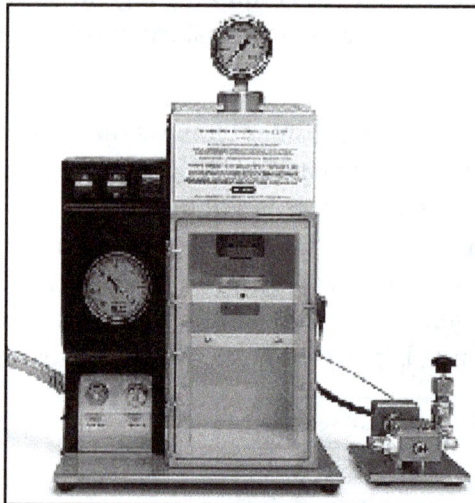

Standard Gene Gun.

Particle bombardment also plays an important role in the transformation of organelles such as chloroplasts, which enables engineering of organelle-encoded herbicide or pesticide resistances in crop plants and to study photosynthetic processes. Limitations to the particle bombardment method relative to Agrobacterium-mediated transformation include frequent integration of multiple copies of the transgene at a single insertion site, rearrangement of the inserted genes, and incorporation of the transgene at multiple insertion sites. These multiple copies can be linked to silencing of the transgene in subsequent progeny. Figures show the different types of gene guns that are currently used in plant transformation.

Generating a whole Transgenic Plant

After innoculation with the agrobacterium, the plant tissue is cultured on media which contains antibiotics to kill the agrobacterium, as well as selective factors. The selective factor is commonly an antibiotic which would normally kill the plant cells. When the gene of interest is added to agrobacterium we add a selective marker- for example an antibiotic resistance gene. This means that only cells in which the foreign DNA has been successfully integrated into their own DNA, will be able to survive in media with the antibiotic.

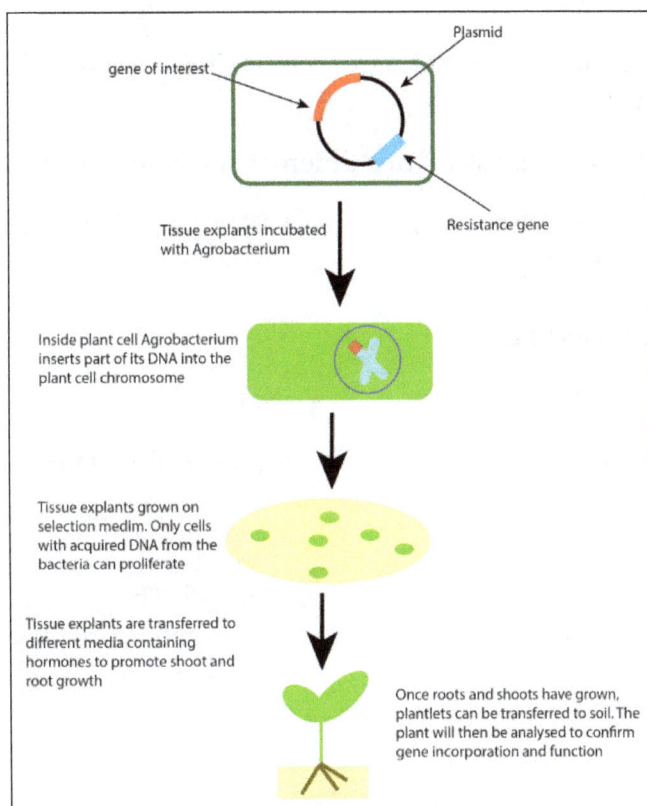

Simplified diagram of Agrobacterium mediated plant transformation.

Plant Transformation Vector

Plant transformation vectors are plasmids that have been specifically designed to facilitate the generation of transgenic plants. The most commonly used plant transformation vectors are termed binary vectors because of their ability to replicate in both E. coli, a common lab bacterium, and Agrobacterium tumefaciens, a bacterium used to insert the recombinant (customized) DNA into plants. Plant Transformation vectors contain three key elements:

- Plasmids Selection (creating a custom circular strand of DNA).
- Plasmids Replication (so that it can be easily worked with).
- Transfer DNA (T-DNA) region (inserting the DNA into the agrobacteria).

Steps in Plant Transformation

- Propagate binary vector in E. coli.

- Isolate binary vector from E. coli and engineer (introduce a foreign gene).

- Re-introduce engineered binary vector into E. coli to amplify.

- Isolate engineered binary vector and introduce into Agrobacteria containing a modified (relatively small) Ti plasmid.

- Infect plant tissue with engineered Agrobacteria (T-DNA containing the foreign gene gets inserted into a plant cell genome).

- In each cell T-DNA gets integrated at a different site in the genome.

There are many variations to these steps. A custom DNA plasmid sequence can be created and replicated in more than one way.

Consequences of the Insertion

- Foreign DNA inserted.

- Insertional mutagenesis (but not lethal for the plant cell – as the organism is diploid).

Problem

We want to transform the whole organism, not just one cell. This is done by transforming plant cells in culture, selecting transformed cells and regenerating an entire plant from the transformed cell (e.g. tobacco).

Plasmid Selection

When the bacteria with the desired, implanted gene are grown, they are made containing a selector. A selector is a way to isolate and distinguish the desired cells. A gene that makes the cells resistant to an antibiotic such as the antibiotics kanamycin, ampicillin, spectinomycin or tetracyclin, is an easy selector to use. The desired cells (along with any other organisms growing within the culture) can be treated with an antibiotic, allowing the desired cells to survive while other organisms cannot. The antibiotic gene is not usually transferred to the plant cell but remains within the bacterial cell.

Plasmids Replication

Plasmids replicate to produce many plasmid molecules in each host bacterial cell. The number of copies of each plasmid in a bacterial cell is determined by the replication origin. This is the position within the plasmids molecule where DNA replication is initiated. Most binary vectors have a higher number of plasmids copies when they replicate in E. coli, the plasmid copy-number is usually less when the plasmid is resident within Agrobacterium tumefaciens. Plasmids can also be replicated in the polymerase chain reaction (PCR).

T-DNA Region

T-DNA contains two types of genes: the oncogenic genes, encoding for enzymes involved in the synthesis of auxins and cytokinins and responsible for tumor formation; and the genes encoding for the synthesis of opines. These compounds, produced by condensation between amino acids and sugars, are synthesized and excreted by the crown gall cells and consumed by A. tumefaciens as carbon and nitrogen sources. Outside the T-DNA, are located the genes for the opine catabolism, the genes involved in the process of T-DNA transfer from the bacterium to the plant cell and the genes involved in bacterium-bacterium plasmid conjugative transfer. The T-DNA fragment is flanked by 25-bp direct repeats, which act as a cis element signal for the transfer apparatus. The process of T-DNA transfer is mediated by the cooperative action of proteins encoded by genes determined in the Ti plasmid virulence region (vir genes) and in the bacterial chromosome. The Ti plasmid also contains the genes for opine catabolism produced by the crown gall cells, and regions for conjugative transfer and for its own integrity and stability. The 30 kb virulence (vir) region is a regulon organized in six operons that are essential for the T-DNA transfer (virA, virB, virD, and virG) or for the increasing of transfer efficiency (virC and virE). Different chromosomal-determined genetic elements have shown their functional role in the attachment of A. tumefaciens to the plant cell and bacterial colonization: the loci chvA and chvB, involved in the synthesis and excretion of the b -1,2 glucan; the chvE required for the sugar enhancement of vir genes induction and bacterial chemotaxis, Cangelosi et al.; the cel locus, responsible for the synthesis of cellulose fibrils; the pscA (exoC) locus, playing its role in the synthesis of both cyclic glucan and acid succinoglycan; and the att locus, which is involved in the cell surface proteins.

Hairy Root Culture

Plant remains major source of pharmaceuticals and fine chemicals and cell cultures have been viewed as promising alternatives to whole plant extraction for obtaining valuable chemicals. The major constraint with the cell culture is that they are genetically unstable and tend to produce low yield of secondary metabolites. A new method for enhancing secondary metabolite production is by transformation of cells or tissues using the natural vector system. Agrobacterium rhizogenes, the causative agent of hairy root disease, is a soil dwelling gram negative bacterium capable of entering a plant through a wound and causing a proliferation of secondary roots. The mechanism of transformation is elaborated in figure. The biosynthetic capacity of the hairy root cultures is equivalent or sometimes more to the corresponding plant roots. Therefore, hairy root cultures have been developed as an alternate source for the production of root biomass and to obtain root derived compounds.

Establishment of Hairy Root Cultures

For the production of hairy root cultures, the explant material is inoculated with a suspension of A. rhizogenes. The bacterial suspension is generated by growing bacteria in Yeast Mannitol Broth (YMB) medium for 2 days at 25 °C under shaking conditions. Thereafter, pelleting by centrifugation (5 x 10 rpm; 20 min) and resuspending the bacteria in YMB medium to form a thick suspension (approx. 10^{10} viable bacteria/ml). Transformation may be inducesd in aseptic seedlings or surface sterilized detached leaves, leaf-discs, petioles, stem segments, from greenhouse grown plants by scratching the leaf midrib or the stem of a plantlet with the needle of a hypodermic syringe containing a small (about 5-10 ul) droplet of thick bacterial suspension of A. rhizogenes.

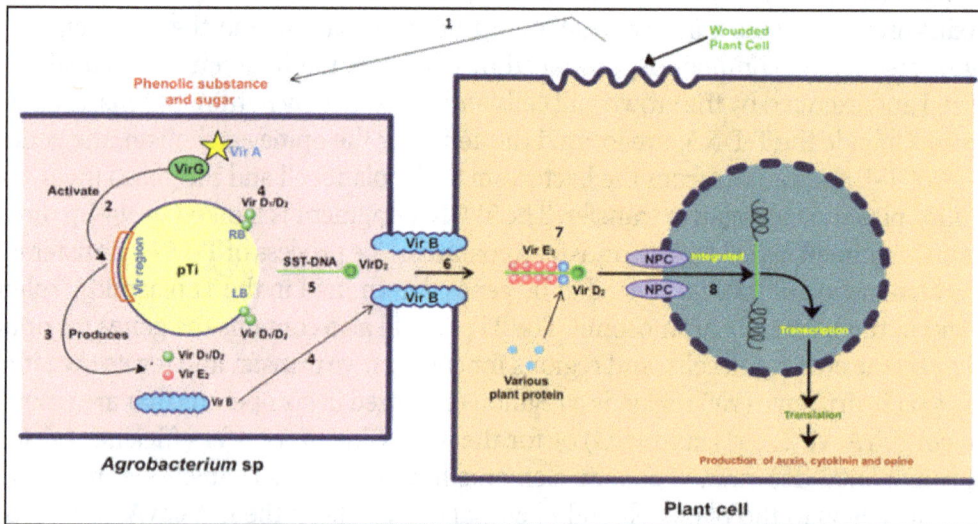

The Agrobacterium injects a plasmid (naked circular DNA) into the host cells.

Wounded plant cell releases phenolic substances and sugar (1); which are sensed by Vir A, Vir A activates Vir G, Vir G induced for expression of Vir gene of Ri-plasmid (2); Vir gene produces all the Vir-protein (3); Vir D_1 and Vir D_2 are involved in ssT-DNA production from Ri-plasmid and its export (4) and (5); the ssT-DNA (associated with Vir D_1 and Vir D_2) with Vir E_2 are exported through transfer apparatus Vir B (6); in plant cell, T-DNA coated with Vir E_2 (7); various plant proteins influence the transfer of T-DNA + Vir D_1 + Vir D_2 + Vir E_2 complex and integration of T-DNA to plant nuclear DNA(8). (LB= left border; RB= Right border; pRi = Ri plasmid, NPC = nuclear pore complex).

Genetics of Transformation

Ri plasmids contain one or two regions of T-DNA and a Vir (Virulence) region, all of which are necessary for tumorigenesis. The Ri plasmid is very similar to Ti plasmid except that their T-DNAs have homology only for auxin and opine synthesis sequences. The T-DNA of Ri plasmid lacks genes for cytokinin synthesis. The T-regions of Ti and Ri plasmids contain oncogenes that are expressed in the plants. Another type, present in Ri plasmids only, appears to impose a high hormone sensitivity on the infected tissue. The T-DNA of Ri plasmids codes for at least three genes that each can induce root formation, and that together cause hairy root formation from plant tissue. Current results indicate that the products of these genes induce a potential for increased auxin sensitivity that is expressed when the transformed cells are subjected to a certain level of auxin. After this stage the transformed roots can be grown in culture without exogenous supply of hormones.

The Ri-plasmids are classified into two main classes according to the opines formed in transformed roots. First, agropine-type strains induce roots to synthesise agropine, mannopine and the related acids. Second, mannopine-type strains which induce roots to produce mannopine and the related acids. The agropine-type Ri-plasmids are very similar as a group and a quite distinct group from the mannopine-type plasmids. Perhaps the most studied Ri-plasmids are agropine-type strains, which are considered to be the most virulent and, therefore, more often used in the establishment of hairy root cultures.

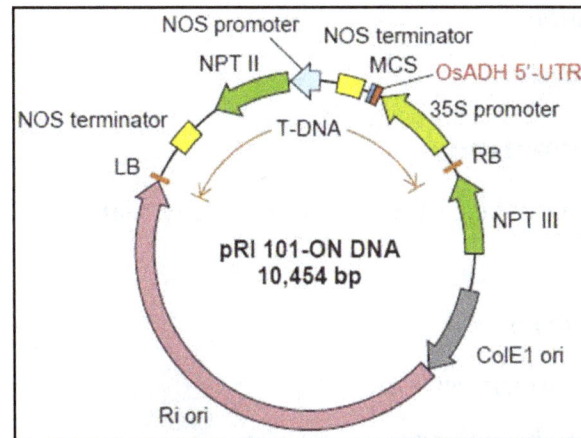

Structure of Ri-plasmid.

The Genes Responsible for Hairy Root Formation

The agropine-type Ri-plasmid consists of two separate T-DNA regions known as the TL-DNA and TR-DNA. Each of the T-DNA fragments is separated from each other by at least 15 kb of non-integrated plasmid DNA. These two fragments can be transferred separately during the infection procedure. The TR-DNA of the agropine type Ri-plasmid carries genes encoding auxin synthesis (tms 1 and tms 2) and agropine synthesis (ags). The mannopine type Ri-plasmids contain only one T-DNA. TL-DNA region consists of four root locus (rol) genetic loci, rol A, rol B, rol C, and rol D, which affect hairy root induction. In particular, rol B seems to be the most important in the differentiation process of transformed cells and also function as induction of hairy roots by hydrolyzing bound auxins leading to an increase in the intracellular levels of indole-3-acetic acid. Gene rol A involved in development of hairy root morphology, rol B is responsible for protruding stigmas and reduced length of stamens; rol C causes internode shortening and reduced apical dominance.

Factors Influencing the Transformation

Following factors influence the transformation process:

- Virulence of A. rhizogenes strains,
- Medium,
- Age of the explant,
- Nature of the explant.

Confirmation of Transformation

Confirmation of transformation can be performed on the basis of following markers:

- Biochemical markers,
- Opines,
- Mannopines,
- Genetic markers,

- Southern hybridization,
- Polymerase chain reaction.

Screening of Transformation

Screening of transformation can be performed by GUS assay, leaf callus assay, rooting and bleaching assays.

Properties of Hairy Roots

Hairy roots have following properties:

- High degree of lateral branching.
- Profusion of root hairs.
- Absence of geotropism.
- They have high growth rates in culture, due to their extensive branching, resulting in the presence of many meristems.
- They do not require conditioning of the medium.

Hairy Roots are Genetically Stable

Hairy roots are genetically stable consequently they exhibit biochemical stability that leads to stable and high-level production of secondary metabolites. Hairy root cultures apparently retain diploidy in all species so far studied. The stable production of hairy root cultures is dependent on the maintenance of organized states. The factors which promote disorganization and callus formation depress secondary metabolite production. The productivity of hairy root cultures is stable over many generations in contrast to disorganized cell cultures. This stability is reflected in both the growth rate and the level pattern of secondary metabolite production.

Application of Hairy Root Cultures

Production of Secondary Metabolites

The hairy root system is stable and highly productive under hormone-free culture conditions. The fast growth, low doubling time, easy maintenance, and ability to synthesize a range of chemical compounds of hairy root cultures gives additional advantages as continuous sources for the production of plant secondary metabolites. Usually root cultures require an exogenous phytohormone supply and grow very gradually, resulting in the poor or insignificant synthesis of secondary metabolites. Hairy roots are also a valuable source of photochemical that is useful as pharmaceuticals, cosmetics, and food additives. These roots synthesize more than a single metabolite; prove economical for commercial production purposes. Many medicinal plants have been transformed successfully by A. rhizogenes and the hairy roots induced show a relatively high productivity of secondary metabolites, which are important pharmaceutical products. Sevon has summarized the most important alkaloids produced by hairy roots, including Atropa belladonna L., Catharanthus trichophyllus L., and Datura candida L.

Metabolic engineering offers new perspectives for improving the production of secondary metabolites by the over expression of single genes. This approach may lead to an increase of some enzymes involved in metabolism and, consequently, results in the accumulation of the target products. This method utilizes the foreign genes that encode enzyme activities not normally present in a plant. This may cause the modification of plant metabolic pathways. Two direct repeats of a bacterial lysine decarboxylase gene, expressed in the hairy roots of Nicotiana tabacum, have markedly increased the production of cadaverine and anabasine. The production of anthraquinone and alizarin in hairy roots of Rubia peregrina L. was enhanced by the introduction of isochorismate synthase. Catharanthus roseus hairy roots harboring hamster 3-hydroxy-3-methylglutaryl coenzyme A reductase (HMGR) cDNA without the membrane-binding domain were found to produce more ajmalicine and catharanthine or serpentine and campesterol than the control.

Production of Compounds not Found in Untransformed Roots

Transformation may affect the metabolic pathway and produce new compounds that cannot be produced normally in untransformed roots. For example, the transformed hairy roots of Scutellaria baicalensis Georgi accumulated glucoside conjugates of flavonoids instead of the glucose conjugates accumulated in untransformed roots.

Changing Composition of Metabolites

Bavage et al. reported the expression of an Antirrhinum dihydroflavonol reductase gene which resulted in changes in condensed tannin structure and its accumulation in root cultures of L. corniculatus. The analysis of selected root culture lines indicated the alteration of monomer levels during growth and development without changes in composition.

Table: Pharmaceutical products produced using hairy root cultures.

Plant species	Product
Bidens spp.	Polyacetylenes
Cinchona ledgeriana	Quinoline alkaloids
Datura spp.	Tropane
Cassia spp.	Anthraquinones
Echinaceapurpurea	Alkaloids

Ti and Ri Plasmids

Agrobacterium species harboring tumor-inducing (Ti) or hairy root-inducing (Ri) plasmids cause crown gall or hairy root diseases, respectively in plants. Agrobacterium tumefaciens is a plant pathogen that induces tumor on a wide variety of dicotyledonous plants and the disease is caused by tumor-inducing plasmid (pTi). Similarly Agrobacterium rhizogenes is a plant pathogen that induces hairy roots on a wide variety of dicotyledonous plants and the disease is caused by root-inducing plasmid (pRi). Virulence (vir) genes of Ri as well as of Ti plasmids are essential for the T-DNA transfer into plant chromosomes. These natural plasmids provide the basis for vectors to make transgenic plants. The plasmids are approximately 200 kbp in size. Both pTi and pRi are unique in two respects: (i) they contain some genes, located within their T-DNA, which have regulatory sequences recognized by plant cells, while their remaining genes have prokaryotic regulatory sequences, (ii) both plasmids naturally transfer a part of their DNA, the T-DNA, into the host genome, which makes Agrobacterium a natural genetic engineer.

Complete sequence analysis confirms that the pathogenic plasmids contain gene clusters for DNA replication, virulence, T-DNA, opine utilization and conjugation. T-DNA genes have lower G + C content, which is presumably suitable for expression in host plant cells. Besides these genes, each plasmid has a large number of unique genes. Even plasmids of the same opine type differ considerably in gene content and are highly chimeric in structures. The plasmids seem to interact with each other and with plasmids of other members of the Rhizobiaceae and are likely to shuffle genes of infection between Ti and Ri plasmids. Plasmid stability genes are talked about, which are important for plasmid evolution and construction of useful strains.

The Ti Plasmid

The Ti plasmid contains all the genes which required for tumor formation. Virulence genes (vir-genes) are also located on the Ti plasmid. The vir genes encode a set of proteins responsible for the excision, transfer and integration of the T-DNA into the plant nuclear genome.

The basic elements of the vectors designed for Agrobacterium-mediated transformation that were taken from the native Ti-plasmid are:

- The T-DNA border sequences, at least the right border, which initiates the integration of the T-DNA region into the plant genome,

- The vir genes , which are required for transfer of the T-DNA region to the plant, and

- A modified T-DNA region of the Ti plasmid, in which the genes responsible for tumor formation are removed by genetic engineering and replaced by foreign genes of diverse origin, e.g., from plants, bacteria, virus. When these genes are removed, transformed plant tissues or cells regenerate into normal-appearing plants and, in most cases, fertile plants.

The T-DNA region genes are responsible for the tumorigenic process. Some of them control the production of plant growth hormones that cause proliferation of the transformed plant cells. The T-DNA region is flanked at both ends by 24 base pairs (bp) direct repeat border sequence called T-DNA borders. The T-DNA left border is not essential, but the right border is indispensable for T-DNA transfer. Ti plasmid is grouped into two general categories:

- Nopaline type pTi

- Octopine type pTi

Both types of plasmid are shown in figure.

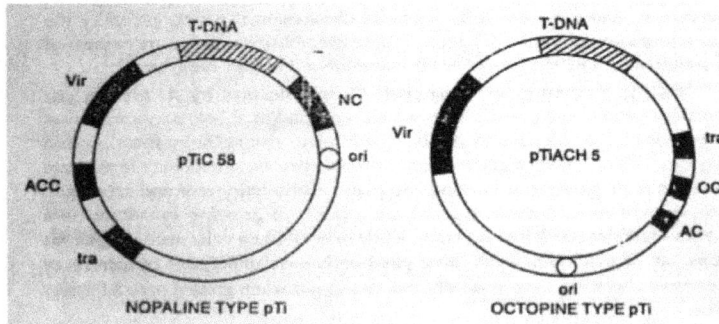

The Ti plasmid: (A) nopaline type pTi; (B) Octopine type pTi.

Ri Plasmid

Agrobacterium rhizogenes is a soil born gram negative bacterium. It causes hairy root disease of many dicotyledonous plants. The ability of A. rhizogenes to incite hairy root disease is confirmed by a virulence plasmid, which is similar to that found in Agrobacterium tumefaciens which causes Crown gall tumors of plants. The virulence plasmid of A. rhizogenes is commonly known as the Ri-plasmid (pRi). The pRi have extensive functional homology with the pTi. The pRi contains distinct segment(s) of DNA, which is transferred to plant genome during infection. The transfer T-DNA to the plant genome is mediated by another segment on the plasmid known as the virulence (vir) region. All strains of A. rhizogenes are known to produce agrocinopine.

Role of Ti Plasmid in Genetic Engineering of Plants

Genetic engineering of plants is carried out by introducing DNA into a cell in culture that can grow into a mature plant. An efficient vector for introduction of recombinant DNA into plant cells has been developed from plant viruses. Furthermore, the Ti plasmid isolated from the bacterium Agrobacterium tumefaciens also serves as a vector for inserting foreign DNA.

Agrobacterium tumefaciens is a gram negative, soil bacterium and a plant pathogen that induces tumour-like growths on plants called crown gall tumours. Gene transfer from the bacterium to the plant occurs naturally, resulting in tumours. Tumours can also be induced in gymnosperms and dicotyledonous angiosperms by inoculation of wound sites with A. tumefaciens.

Evidence suggests that crown gall tissue represents true oncogenic transformation because the undifferentiated cell mass of the tumour (callus) can be cultured in vitro even if bacteria are killed by antibiotics, still retaining its tumour-like properties. These properties include unlimited growth as a callus, and synthesis of opines, such as octopine and nopaline which are unusual amino acid derivatives not present in normal plant tissue.

The metabolism of opines is a central feature of crown gall disease. Plant cells acquire the property of opine synthesis when they are colonised by A. tumefaciens. The bacterium utilizes opine as its sole source of nitrogen and carbon.

The virulent strains of A. tumefaciens contain a Ti plasmid that confers tumour-inducing properties on the bacterium. Earlier investigators on crown gall tumours had observed that continued presence of Agrobacterium is not required to maintain plant cells in their transformed state.

These plasmids were known as tumour-inducing plasmids (Ti plasmids). Ti plasmids also contain information about the specific type of opine that is synthesised in the transformed tissue and utilised by the bacterium.

Furthermore, the complete Ti plasmid is not found in the plant tumour cells. A specific segment of the plasmid, about 23 base pair in size is found integrated in plant nuclear DNA at a random site. This segment of DNA transferred from the plasmid is called T-DNA (transferred DNA).

T-DNA carries genes responsible for conferring unlimited growth and ability to synthesise opines upon the transformed plant tissue. The genes responsible for T-DNA transfer are located in a separate part of the Ti plasmid called vir (virulence) segment.

Two of these genes (virA and virG) are expressed constitutively at a low level of control. The vir gene expresses a protein that forms a conjugative plasmid through which T-DNA is transferred to the nucleus. Once inside the nucleus, T-DNA is incorporated randomly into nuclear DNA by a process of illegitimate recombination. In addition to plants, Agrobacterium can transfer DNA to other bacteria, yeasts and filamentous fungi.

Thus, Ti plasmid serves as a natural vector in genetic engineering of plant cells because it can transfer its T-DNA from the bacterium to the plant genome. The wild type Ti plasmids, however, are not suitable as vectors because of the presence of oncogenes in T-DNA which result in dis-organised growth in recipient plant cells.

To accomplish efficient regeneration in plants, attempts have been made to delete all of the oncogenes from the plasmid. Indeed, when Agrobacterium carrying non-oncogenic plasmids were allowed to transfer the modified T-DNA to plant cells, no tumours were produced.

Transgenic Plants from Transformed Protoplasts

Somewhat parallel to transfection in animal cells, protoplast transformation has been achieved in plants. Plant protoplasts from which the rigid cell wall has been removed, are induced to take up DNA. A number of chemicals, in particular polyethylene glycol promote gene transfer across the protoplast membrane. Alternatively, DNA uptake can be induced by electroporation.

In a small proportion of the protoplasts, the DNA is incorporated stably into the genome (transformed protoplasts). The first successful experiments on plant regeneration using protoplasts were carried out on tobacco and petunia, subsequently in monocots such as Lolium.

Gene Silencing

The term gene silencing is commonly used to describe the "switching off" of a gene by a mechanism without genetic modification. The term gene silencing refers to an epigenetic phenomenon, the heritable inactivation of gene expression that does not involve any changes to the deoxyribonucleic acid (DNA) sequence. While this phenomenon has initially been studied in transgenic plants, its relevance in the regulation of endogenous plant genes has become increasingly apparent. Below some cellular components are mentioned where gene silencing occurred:

- Chromatin and heterochromatin
- Dicer
- dsRNA
- Histones
- MicroRNA
- siRNA
- Transposons

Gene silencing has following two major subdivisions by which genes are regulated:

- Transcriptional gene silencing (TGS) and
- Posttranscriptional gene silencing (PTGS).

Transcriptional Gene Silencing (TGS)

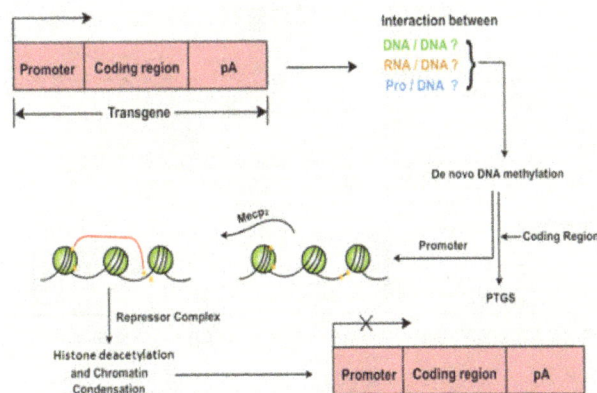

Schematic representation of DNA methylation – mediated transcriptional gene silencing (TGS).

Transcriptional gene silencing is the product of chromosomal histone modifications, creating an environment of heterochromatin, which is surrounded to a gene that makes it inaccessible to transcriptional machinery (RNA polymerase, transcription factors, etc.). TGS blocks primary transcription from nuclear DNA and is in most cases associated with DNA methylation and chromatin condensation in nearly all organisms that possess a DNA methylation system.

Post-transcriptional Gene Silencing (PTGS)

Post-transcriptional gene silencing is the product of transcribed mRNA of a specific gene being silenced. When mRNA was destructed, then translation to form an active gene product (in most cases, a protein) will be prevented. A general process of post-transcriptional gene silencing is by RNAi. PTGS involves a cytoplasmic, target sequence-specific RNA degradation process that is possibly activated by double-stranded RNA (dsRNA). This dsRNA is independent of ongoing translation.

TGS can be transmitted generation to generation by meiosis whereas PTGS is usually lost during meiosis. In PTGS, double stranded RNA is interred into a cell and gets chopped up by the enzyme known as dicer to form siRNA. siRNA then binds to the RNA-induced silencing complex (RISC) and is unwound. The anitsense RNA complexed with RISC protein and binds to its corresponding mRNA, which is then cleaved by the enzyme slicer rendering it inactive.

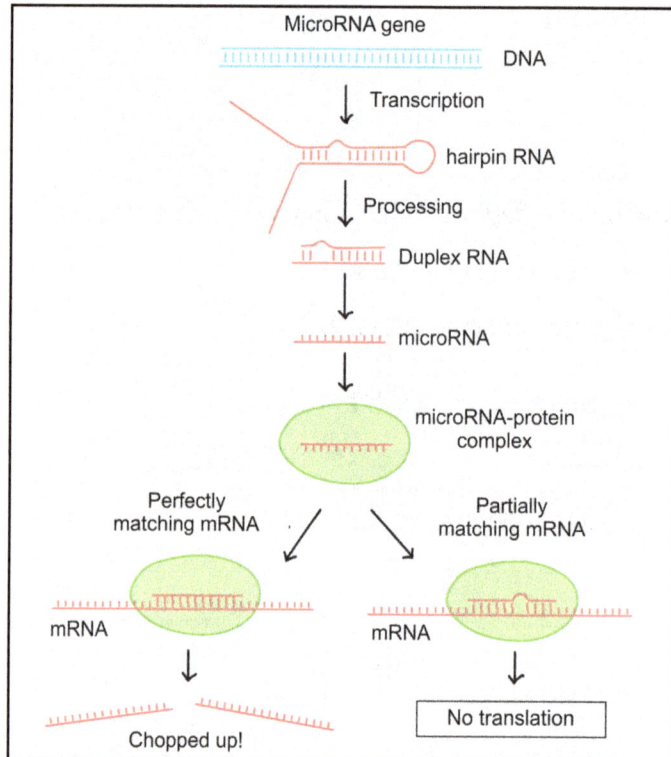

Schematic representation of post translational gene silencing (PTGS).

Binary Vector

Binary vector was developed by Hoekma et al and Bevan. It utilizes the trans- acting functions of the vir genes of the Ti-plasmid and can act on any T-DNA sequence present in the same cell. Binary vector contains transfer apparatus (the vir genes) and the disarmed T-DNA containing the transgene on separate plasmids.

Advantages of Binary Vector

- Small size due to the absence of border sequences needed to define T-DNA region and vir region.

- Ease of manipulation.

Binary Vector System

A plasmid carrying T-DNA containing LB and RB, called mini-Ti or micro-Ti can be sub-cloned in a small E. coli plasmid for ease of manipulation.

The T-DNA of mini-Ti can be introduced into an Agrobacterium strain carrying a Ti plasmid from which the T-DNA has been removed but contains vir region. The vir genes function in trans, causing transfer of the recombinant T-DNA to the plant genome.

The T-DNA plasmid can be introduced into Agrobacterium by triparental mating or by a more simple transformation procedure, such as electroporation.

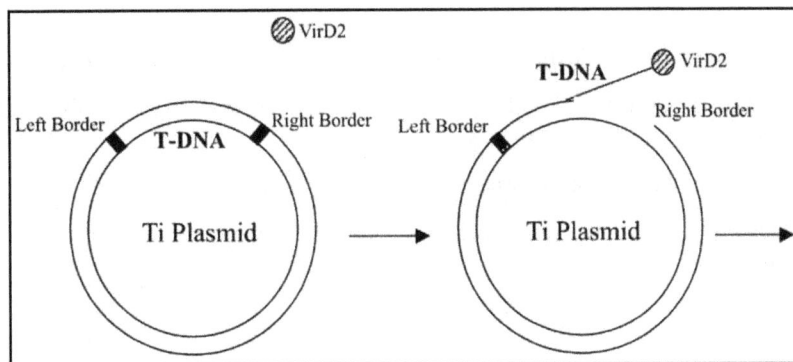

A binary vector system (a) A plasmid containing vir region but no T-DNA, therefore no T-DNA transfer takes place in plant genome. (b) Another plasmid containing T-DNA with Right border (RB) and Left border (LB) but no vir genes. Vir function is supplied in trans by former plasmid.

In Ti-plasmid transformation system, the T-DNA is maintained on a shuttle vector with a broad host range origin of replication, such as RK2 (which functions in both A. tumefaciens and E. coli), or separate origin for each species.

An independently replicating vector is advantageous because maintenance of the T-DNA does not rely on recombination. The copy number of binary vector is not determined by the Ti plasmid, making the identification of transformants much easier.

All the conveniences of bacterial cloning plasmids are incorporated into binary vectors, such as multiple unique restriction sites in the T-DNA region to facilitate subcloning, the lacZ gene for blue–white screening and a λ cos site for preparing cosmid libraries.

Examples of Binary Vector System

- pBIN19-one of the first binary vectors developed in 1980s and was widely used.

- pGreen-A newly developed vector with advanced features than pBIN19.

Both the vectors contain Lac Z gene for blue-white screening of recombinants. The reduction of size of pGreen is due to the presence of pSa origin of replication. An essential replicase gene is housed on a second plasmid, called pSoup which functions in trans. All conjugation functions have also been removed, so this plasmid can only be introduced into Agrobacterium.

Table: Comparison between Binary vectors pBIN19 and pGreen.

Features	pBIN19	pGreen
Size	Large (11777bp)	Small (<5bp)
Position of Selectable marker	Situated near RB. Due to origination of transfer from RB, transfer of selectablemarker before gene of interest will result in plants expressing selectable marker but not with transgene or truncated version.	Situated next to LB
Restriction sites	Limited	Much larger MCS with many restriction sites.

References

- What-is-plant-transformation: scientisterica.wordpress.com, Retrieved 23 June, 2019

- Plant-transformation-using-particle-bombardment, process-of-developing-genetically-modified-gm-crops, bio-technology: nepad-abne.net, Retrieved 15 May, 2019

- What-is-plant-transformation: scientisterica.wordpress.com, Retrieved 20 July, 2019

- Role-of-ti-plasmid-in-genetic-engineering-of-plants, genetic-engineering: biologydiscussion.com, Retrieved 25 August, 2019

Chapter 6
Applications of Plant Biotechnology

Plant biotechnology is used in a variety of different areas such as improving the nutritional quality of different food crops, improving the flavor and texture of fruits and vegetables, and in the process of brewing and malting. It is also involved in the production of industrial enzymes. All these diverse concepts and processes related to plant biotechnology have been carefully analyzed in this chapter.

Genetic engineering of plants provides an opportunity to alter their properties or performance in order to improve upon their utility. Such technology may be used to modify the expression of genes already present in the plants, or to introduce new genes of other species with which the plant cannot be bred conventionally. Thus, it gives greater efficiency to the fulfillment of conventional breeding purposes.

One of the significant applications of such techniques lies in adding single genes to desirable plant types. Plant transformation can be used to introduce new or novel characteristics that create a new market or displace conventional products. The improvement may relate to the nutritional value of the plant or the functional properties in processing or even consumption per se.

Above all, this technology broadens the possibilities of transferring genes between unrelated organisms, and thus creates novel genetic information by specific alteration of cloned genes.

Food Quality

Nutritional Quality

Seed crops play an important role in human and animal nutrition. Just a few cereals contribute to nearly fifty per cent of the total food calories. Similarly, seven species of grain legume account for a large portion of our calorie intake.

However, cereals and legumes contain certain proteins that are deficient in amino acids like lysine and threonine. Legumes are also deficient in sulphur amino acids. Some other seed crops like rice offer a better balance of amino acids, but fall-out on their overall protein levels.

Common logic follows that each of these foods could be catapulted to perfection if their deficiencies could be overcome by borrowing those missing traits from other crops. That is exactly what plant biotechnology does – the transfer of single or multiple genes to plants lacking important components.

Recently, Professor Ingo Potrykus at the Swiss Federal Institute of Technology (Zurich) and Dr Peter Beyer of the University of Freiburg (Germany) developed the 'Golden Rice', which has greater levels of pro-vitamin A or b-carotene.

This modified rice is expected to provide nutritional benefits to those suffering from vitamin-A deficiency related diseases, including irreversible blindness in hundreds of thousands of children annually. Adequate vitamin-A content can also reduce the mortality associated with infectious diseases such as diarrhea and childhood measles by enhancing the activity of the human immune system.

Genetic tools can be used to alter the carbohydrate, fat, fibre and vitamin content of food. Another useful application is that of picking up genes from protein-rich cereals and transferring them to low-protein food.

Transgenic tools are also being used to improve the nutritional value of crops by reducing their anti-nutritional factors (like protease inhibitors and haemaglutinins in legumes). Problems associated with the flatulence in certain foods can also be addressed by manipulating the dietary fibre and oligosaccharide content.

Biotechnical applications are extremely useful in case of wheat as well. The quality of wheat is determined by the presence of seed-storage proteins of the grain. Thus, its quality may be improved by manipulating the presence of these proteins. More gluten proteins can also be added to give enhanced elasticity to dough. Further, the starch content of wheat can be altered to suit the properties of products like noodles.

Functional Quality

Transformation can be applied to fruits and vegetables to improve their flavour and texture by manipulating their maturing process. The performance of plant products during their processing can also be improved by genetic engineering. For instance, the first genetically engineered food, the Flavr-Savr tomato was genetically manipulated to slow down its ripening, and has a longer shelf-life.

Another common strategy to control ripening is to curb the production of the ripening hormone ethylene. Ethylene is produced from S-adenosylmethionine by conversion to 1-amino-cyclopropane- 1-Carboxylic acid (ACC) in the presence of ACC Synthase, followed by the generation of ethylene by an ACC oxidase or ethylene-forming enzyme.

Ripening can be delayed by directing antisense constructs against either of these enzymes, or by removing ACC with an ACC deaminase. Fruits may then be ripened as required by exposure to an artificial source of ethylene.

Transgenic tomato with delayed ripening.

Malting and Brewing

The production of beer involves the germination of barley under controlled conditions. The quality of beer thus depends largely upon the composition of the barley grain. Many qualities of these grains can be substantially improved through genetic engineering. For instance, improving the stability of the barley enzymes (especially at high temperatures) can enhance its effectiveness at the temperature used during mashing. The flavour of the beer can also be manipulated by genetically treating the barley. One such technique is that of reducing the levels of lipo-oxygenase.

Storage Carbohydrates

Increasing the levels of certain enzymes like ADP pyrophosphorylase can enhance the starch synthesis of food products. This can improve the yields of starchy foods. Transformation can also alter the properties of plant starches. The proportion of amylase and amylopectin in the starch and the quality can also be regulated. This would allow the tailoring of starch to meet requirements for specific foods or industrial products.

Transgenic plants with increased levels of fructans (a form of glucose) are already being produced using a levansucrase from bacteria. The sucrose content of plants can also be manipulated to enhance the quality of sugar crops like sugar cane and sugar beet.

Disease Resistance

Insect Resistance

Genetic engineering has proved to be a boon for producing pest-resistant plants. This technology has overcome the shortcomings of using chemical pesticides. Of late, the technique of introducing disease- resistant genes into plant species has also gained tremendous popularity.

For instance, protease inhibitors can prevent the digestion of proteins by insects, and hence slow down their growth rate. The transfer of such proteins to the plants acts as a natural protection mechanism against insect attack.

Certain bacterial genes have also proved to be quite effective in preventing pest damage. Bacillus thuringiensis (Bt) produces Bt Toxin, which is effective against insect larvae. Transgenic plants harbouring Bt genes have been produced in crops like soya bean, maize and cotton, and have proved to be resistant to pest attacks.

Many other serochemicals (chemicals which alter the insect behaviour) are produced by certain insect and plant species. Transferring these to other plants can be very effective in checking disease incidence. To take another example, the susceptible potato crop does not contain anti-feedant chemicals like farnase, a terpenoid and other related compounds.

These are produced by aphid resistant plant species such as Solanum berthaultii (in leaf hairs). These compounds act by eliciting an attack response in aphids, so that they are unable to establish themselves on the crop. Transferring these genes to the potato crop can protect it from aphid menace.

Virus Resistance

Production of transgenic plants with resistance to viruses is one of the most successful applications of plant transformation. Several strategies involving the expression of the viral genome in the plant have proven effective. For example, the expression of coat protein gene from virus has been widely successful. Both sense and antisense expression of parts of the viral genome can be protective against viral infection.

Nematode Resistance

Novel genes for nematode resistance offer an alternative approach to the production of nematode-resistant plants. Genetic engineering provides an opportunity to develop transgenic plants with genetic resistance to these long-term plant pests, and thus reduce the reliance on chemical nematicides in agriculture.

Herbicide Resistance

The choice of a herbicide is very critical as it carries a high risk of inducing resistance. Weeds may rapidly develop multiple herbicide resistance in some systems when several classes of herbicide act on the same molecular target. Here again, herbicide resistance genes offer protection by detoxifying the herbicide (converting it to an inactive form).

Enhancing Photosynthetic Efficiency

The process of photosynthesis is the most significant mechanism for adding energy to the plants. However, even the most efficient plants can utilise only about three to four per cent of the full sunlight. Biotechnology is now being used to improve the level of photosynthetic efficiency of RuBP-Case (Ribulose bis phosphate carboxylase, involved in carbondioxide fixation).

This enhances the efficiency of catalysis and reduces the competitive oxygenase function (as RuBP Case also behaves as an oxygenase). Useful variants can also be produced by combining the genes coding for large and small sub-units of the enzymes from different species.

Two different ways of doing this are:

Abiotic Stress Tolerance

Plant productivity suffers major losses due to various forms of stress during the course of their development. These stress factors include temperature, salinity, drought, flooding, UV light, and various infections. While the molecular basis of such responses are not yet clear, we know that they include de novo synthesis of specific proteins (under temperature shock) and enzymes (alcohol dehydrogenase under anaerobiosis and phenyl alanine amino lyase under UV irradiation).

The genes responding to abiotic stress have been cloned and sequenced in many laboratories, including that of the authors who identified and transformed a gene encoding glyoxalase 1 to confer tolerance to plants.

The regulatory sequences of some of the genes have also been identified. For instance, the 5' promoter sequence of alcohol dehyhrogenase has been linked to the CAT reporter (Chloremphenicol Acetyl Transferase) gene and transferred to tobacco protoplasts where O_2-sensitive expression has been demonstrated.

Such environmentally inducible promoters will certainly become useful tools to study gene expression, and this work will lay the foundation for the transfer of stress-responsive genes under regulated promoters to susceptible species. Recently tomato plants that are resistant to salinity have been developed.

Genes from varied organisms like marine resources can also be used to improve upon plants in various ways. This is an innovative step towards developing salt tolerant species, by transferring the genes from marine plants (halophytes) to grain and vegetable crops.

Similarly a gene, which encodes a protein from a flounder fish, has been transformed to plants to protect them against freezing damage. This protein could be useful in preventing frost damage in post-harvest storage. Thus, freezing could be used to preserve the texture and flavour of some fruits and vegetables, which are currently not suitable for freezing.

Development of Nitrogen Fixing Capacity in Non-leguminous Crops

While the application of nitrogenous fertilisers has proved to be an efficient route to improving crop yields, it continues to be an expensive proposition. The alternative is to provide a natural source of nitrogen within the plant. Introducing nitrogen-fixing microorganisms can do this.

Such micro-organisms are capable of fixing atmospheric nitrogen in the presence of nitrogen fixing bacteria Rhizobium. Transforming the nitrogen fixing genes (nif genes) from leguminous to non-leguminous crops can offer a cost-effective alternative to the expensive fertilisers.

However, other ways of improving the nitrogen yield in plants can be achieved by increasing the efficiency of fixation process in symbiotic bacteria, increasing the efficiency of fixation process in the synthetic bacteria, modifying the nitrogen-fixing bacteria to maintain nitrogen fixation in presence of exogenous nitrogen.

Cytoplasmic Male Sterility

A lot of research has gone into explaining the mechanism of Cytoplasmic Male Sterility (CMS). This trait results in the production of non-functional pollen in mature plant species like sorghum, maize and sugar beet, and hence facilitates the generation of valuable high yielding hybrid seed.

Cytoplasmic male sterility in these plant species is basically associated with the reorganisation of mitochondrial DNA and the synthesis of novel polypeptides. The rapidly developing biotechnological tools may eventually enable the transfer of the CMS trait to male fertile lines. Genetically engineered male sterility also holds a great potential for generation of hybrids in agriculture.

Plant Development

Development of a plant is a complex process, which involves the role of light receptors like phytochrome, chloroplast gene expression, mitochondrial gene expression in relation to male sterility, storage product accumulation, and storage organ (fruits) development.

It is now possible to clone and sequence various genes responsible for plant development. This has increased the possibility of manipulating the expression of these genes, and subsequently the process in which they are involved. For instance, early flowering genes have been reported to alter the properties of the late maturing varieties.

The isolation of specific promoter elements has also helped design crops that express proteins in specific tissues. Genes responsible for colour formation can be transferred to plants bearing colourless flowers. What's more, manipulation of genes that control flowering and pollen formation can generate transgenic plants with altered fertility. Expression of leafy and APETALAI gene in Arabidopsis has resulted in precocious flowering.

Similarly, the putative hormone receptors in plants influence the sensitivity of different tissues to growth regulators, and their subsequent differentiation and development. The introduction of wild type or modified genes for specific growth regulators has proved effective in manipulating plant development (like changing the maturity time or the number and size of potato tubers). This approach can be applied for modifying flowering response, fruit development and expression of storage protein genes.

Useful Proteins from Plants

Many plants are now being used to produce useful proteins. This has given birth to Neutraceuticals – a word coined for made up food. These foods are also known as functional foods. The neutraceuticals include all 'designer' foods from the vitamin-enriched breakfast cereals to Benecol, a margarine spread that actually lowers LDL cholesterol. A leading American company, Novartis Consumer Health, estimates the US market for functional foods at around ten billion dollars, with an expected annual growth rate of ten per cent.

Vaccine Production from Plants

Plants are a rich source of antigens for the immunization of animals. Transgenic plants may be developed to produce antigenic proteins or other molecules. Production of the antigen in an edible part of the plant could prove to be an easy and effective mode of delivery system for the antigen in an edible part of the plant could prove to be an easy and effective mode of delivery system for the antigen.

Potential applications of this technology would include efficient immunisation of humans and animals against disease and control of animal pests. For instance, antigens for the Hepatitis B virus have been successfully expressed in tobacco plants and used to immunise mice. Mice fed potatoes expressing the P-sub unit of E.coli enterotoxin LT-B have also produced antibodies, thus protecting against the bacterial toxin.

This technique promises to pave the way for inexpensive immunisation against several human diseases. Oral vaccines against cholera have already been expressed in plants. Generation of antigens through plants is not only cost-effective, but can also be mass produced, and easily recovered.

Production of Industrial Enzymes and Biodegradable Plastics

Production of Industrial Enzymes

The first commercialized 'industrial proteins' produced from transgenic plants were avidin and β- glucuronidase, both of which were produced in maize. After this ProdiGene Inc. company went for the large scale production of trypsin, which is difficult to produce in conventional recombinant systems. A selected list of industrial enzymes produced in transgenic plants and their important application is given in table.

Table: Some example of industrial enzymes produced in transgenic plants and their important applications.

Enzyme	Applications
α-Amylase	Food processing
Avidin	In diagnostic kits
Cellulase	Production of alcohol from cellulose
β-glucanase	In brewing industry
β-glucuronidase	In diagnostic kits
Lignin peroxidase	In paper manufacture
Phytase	Improved phosphate utilization
Trypsin	Pharmaceutical
Xylanase	Biomass processing, paper and textile industries

Trypsin

Trypsin is an important proteolytic enzyme and its production by conventional recombinant approaches is rather difficult. This protein is currently harvested from bovine and protein pancreases. It has wide range of applications including the production of pharmaceuticals, such as insulin, vaccine production and wound care. Transgenic maize plants were used for the expression of bovine pancreatic trypsin. Expression levels were much higher for trypsinogen, the inactive precursor for trypsin, than those for active trypsin. Seed preferred expression of the zymogen form yields the highest expression level of a protease in transgenic plants.

Avidin

Avidin was the first commercial transgenic protein produced. Until plant derived avidin entered the market, the source for commercial production of avidin was chicken egg white. Avidin is a medium-sized, glycosylated protein. Recombinant avidin was produced in transgenic maize. Glycosylation of maize-derived avidin is similar to that seen in native avidin although some glycosyl residues are added to recombinant avidin. Avidin system provides a good example of the economic promise of transgenic plants for large scale production of heterologous proteins.

Cellulase and Xylanase

Cellulases and xylanases, which are normally produced by gut microorganisms in ruminants, have been produced in number of different plants. These enzymes are used in the bioethanol, textile, pulp and paper industries and for the production of animal feed. In all these processes, they are basically involved in the degradation of plant material (cellulose). To avoid the risk of autodigestion of plant cells (by these enzymes), engineered, thermostable, forms of the enzymes with high temperature optima were used. Thus, cellulase and xylanase, produced by transgenic plants, are inactive at the temperatures at which plants normally grow. The activity of these enzymes is restored on heating the plant extracts. An interesting point to note is that these enzymes would require only the minimum of purification.

Phytase

Phytase is a hydrolytic enzyme that catalyses the hydrolysis of phytate (inositol hexaphosphate) to inositol and inorganic phosphate. Phytate is present in high quantities in many plant seeds used as a feed to pigs and poultry. These animals do not possess the enzyme phytase; hence they cannot derive the nutrient phosphate from phytate. The undigested phytate gets excreted and accumulates in the soil and water, leading to eutrophication. Transgenic plants capable of synthesizing phytase in their seeds have been developed. These seeds are used in the feed of animals. The phytase enzyme has successfully solved nutritional (phosphate) and environmental (eutrophication) problems.

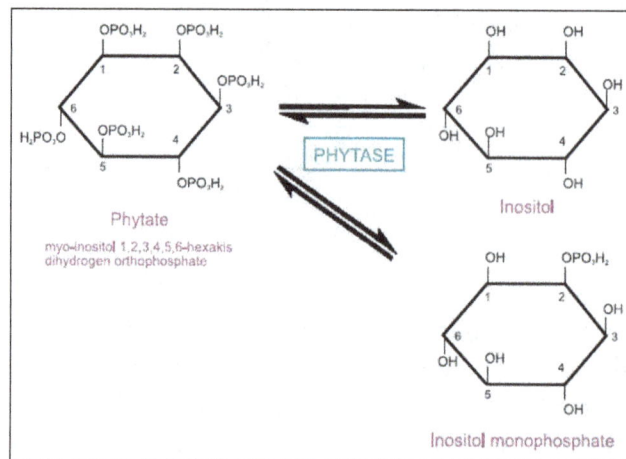

The action of phytase.

Production of Biodegradable Plastics

Biodegradable plastics (Bioplastics) are chemically polyhydroxyalkanoates (PHAs). They are intracellular carbon and energy storage compounds, produced by many microorganisms. They are biodegradable polymers, and are elastic in nature. These compounds are currently produced by microbial fermentation. Several experimental studies are in progress to produce bulk quantities of bioplastics in plants. Among the PHAs polyhydroxy butyrate (PHB) is the most important one.

PHA-chemistry and Properties

PHA serves as lipid reserve material in bacteria. The granules of PHA, stored within the cells are clearly visible under electron microscope. PHAs a re linear polyester polymers composed of hydroxyacid monomers. Structures of PHAs are given in figure. The most commonly found monomers are 3 hydroxy acids with a carbon length ranging from C_3 to C_{14}.

Homopolymer of PHA – Polyhydroxybutyrate

The most common PHA is polyhydroxybutyrate. It is a polyester with 3-hydroxybutyrate as the repeating unit. PHB is a homopolymer PHA. It is hard and inflexible. Being a high molecular weight compound, the accumulation of PHB in huge quantities does not affect the osmotic pressure within the cell. The reserve carbon compound PHB can be oxidized to carbon dioxide and water, releasing large amount of energy. Bacteria need energy to maintain pH gradient and concentration gradient of several compounds. This energy called maintenance energy essential for the survival of cells, is met by the reserve material PHB.

Heteropolymers of PHA

Majority of polyhydroxyalkanoates except PHB contains 2 or more different monomers called as heteropolymers. These heteropolymers are usually composed of a random sequence of monomers in different chains. Beside 3 hydroxy acids several other hydroxyl acids are found in the structures of PHA (eg. 4-hydroxybutyrate). It depends on the organism and the nature of carbon source supplied during accumulation of the polymer. The properties of PHA mostly depend on the nature of the monomers it contains. In general PHA with longer side chains and hetero polymeric PHA are more flexible and soft.

Hydroxybutyrate-co-3-hydroxyvalerate (PHB/V)

By changing the medium composition and selecting a specific organism, the chemical structure of PHA can be altered. In the presence of glucose and propionic acid, the organism Ralstonia eutropha produces a copolymer of 3-hydroxy-butyrate and 3-hydroxyvalerate. The presence of 3-hydroxyvalerate monomers makes PHB/V flexible and stronger. The properties of PHB/V are similar to those of polypropylene, and therefore it is commercially more useful.

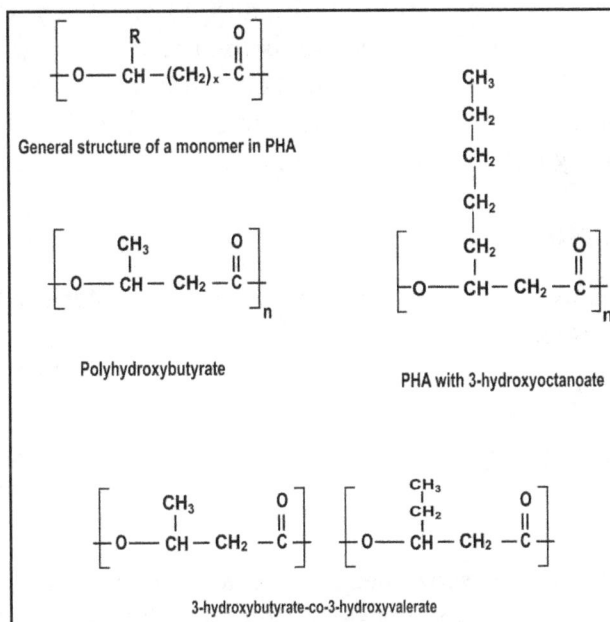

Structures of polyhydroxyalkanoates.

Biosynthesis of PHB/V

Some strains of Ralstonia eutropha are capable of synthesizing polyhydroxybutyrate-co-hydroxy-valerate (PHB/V). For the formation of PHB/V, glucose and propionic acid are required as substrates. Propionyl CoA is responsible for the synthesis of 3-hydroxyvalerate.The three enzymes involved in the synthesis of 3-hydroxy butyrate also participate in the formation of 3-hydroxyvalerate.The polymer PHB/V contains 3-HB and 3-HV monomers in a random sequence. The relative concentrations of glucose and propionic acid in the culture medium determine the chemical composition of PHB/V. The biosynthetic pathways are described in figure.

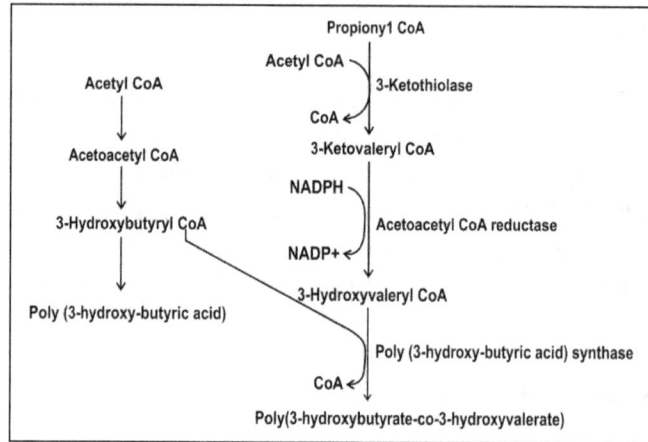

Biosynthesis of PHB/V

Polyhydroxy Butyrate (PHB)

Biosynthesis of PHB

PHB is synthesized in 3 reaction steps starting with acetyl CoA. Acetyl CoA is converted to acetoacetyl CoA by the enzyme 3–ketothiolase which is then reduced to 3- hydroxybutyryl CoA by acetoacetyl CoA reductase. The reducing equivalents are supplied by NADPH. The enzyme PHA synthase is responsible for the addition of 3- hydroxy butyrate residues to the growing PHB chain. Majorly, PHP synthesize in cytoplasm and plastid of cell.

PHB Production in Cytoplasm

Starting from acetyl CoA, polyhydroxy butyrate production is a 3 stage pathway involving the following enzymes (with corresponding genes).

- 3-Ketothiolase (phaA)

- Acetoacetyl-CoA reductase (phaB)

- PHB synthase (phaC)

The three genes coding the respective enzymes have been isolated from Alcaligenes eutrophus and cloned. The cytoplasm of plant cell contains 3-Ketothiolase. Therefore only two genes (phaB and phaC) coding acetoacetyl CoA reductase and PHB synthase were transferred to develop Arabidopsis . By this approach the quantity of PHB produced was very low.

PHB Production in Plastid

In this case all three genes (phaA, PhaB, PhaC) of PHB synthesis was separately fused with a coding sequence of transit peptide bound to N- terminal fragment of Rubisco (ribulose

1,5-bisphosphate carboxylase oxygenase) subunit protein. These genes were then directed to chloroplast. The genes expression was carried out by CaMV 35S promoter. Transgenic Arabidopsis plants with each gene construct were first developed. Then a series of sexual crossings were carried out between the individual transformants. The transgenic plants developed by this approach yielded good quantity of bioplastics and there was no adverse effect on the growth and fertility of these plants.

Production of Bioplastics in Cotton Fibers

Cotton fibres contain the enzyme β-ketothiolase. Therefore the genes for the other two enzymes of PHB pathway (phaB and phaC) from Alcaligenes eutrophus were transferred into meristems of cotton plant by particle bombardment. In this case large quantities of PHB were produced in the fibres of transgenic cotton plants.

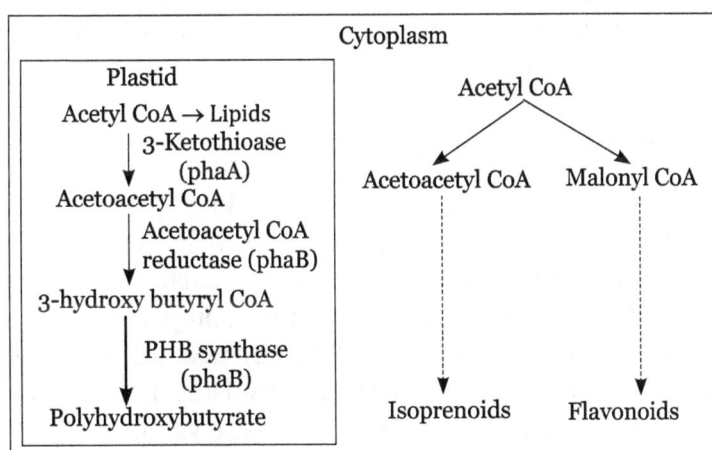

Chloroplast synthesis of PHB.

Production of PHA Copolymers

The other bioplastic, composed of polyhydroxyalkanoate copolymer is a polymer made up of longer monomers. It is less crystalline and more flexible compared to PHB. The PHAs are produced from the intermediates of β oxidation of fatty acids like 3-hydroxy acyl-CoA. PHAs have also been produced through genetic manipulations of peroxisomes and glyoxisomes.

Applications of PHAs

There are several applications for PHA produced by micro-organisms within the medical and pharmaceutical industries, chiefly due to their biodegradable properties:

- In Fixation and orthopaedic applications, including sutures, suture fasteners, meniscus repair devices, rivets , tacks, staples, screws (including interference screws).

- It also used in making of bone plates and bone plating systems, surgical mesh, repair patches, slings, cardiovascular patches, orthopedic pins (including bone.lling augmentation material), adhesion barriers, stents, guided tissue repair/regeneration devices, articular cartilage repair devices, nerve guides, tendon repair devices.

- PHAs can also be useful in making of skin substitutes, dural substitutes, bone graft substitutes, bone dowels, wound dressings, and hemostats.

- PHB can be implanted in the human body without rejection. This is because PHB does not produce any immune response and, thus, it is biocompatible. PHB has several medical applications like, durable bone implants and wound dressing.

Edible Vaccines

Edible vaccines vaccination is a disease preventive measure, where the immune system of a person is boosted against a particular disease. The introduction of a foreign DNA into the eukararyotic cells is called transfection. And the eukaryotic organism which has taken the DNA is termed as transgenic. Transgenic plants are used as recombinant protein production systems and the edible plant tissue functions as an oral vaccine.

Benefits of Edible Vaccine

An edible vaccine in contrast to the traditional vaccines would not require elaborate production facilities, purification, strerilization, and packaging or specialized delivery systems. Injectable vaccines are usually made by synthesizing antigenic proteins in mammalian cell culture. The process is very expensive, requiring speciallybuilt manufacturing facilities. Complicated and time consuming procedures are necessary to purify the proteins from cell cultures. Additionally, vaccines synthesized in mammalian cells can potentially be contaminated with organisms that are pathogenic to humans - a problem that would not arise when using plants to synthesize vaccines. Fruits and vegetables carrying vaccines are also advantageous in that they can be delivered without needles, do not require refrigeration and can be made, less expensively, right in the area in which they will be delivered.

When edible vaccine is taken orally they stimulate the immune system to generate antibodies to a pathogen. The immune system keeps a record of this first encounter and, if the body is later infected with the intact pathogen, the immune system is able to mount a stronger, more effective response.

Potatoes, tomatoes, rice tobacco, lettuce, safflowers, carrots and other plants have been genetically engineered to produce insulin and certain vaccines. If future clinical trials prove successful, the advantages of edible vaccines would be enormous, especially for developing countries. The transgenic plants may be grown locally and cheaply.

Transforming Plants into Vaccine Factories

The process of creating a plant that produces a vaccine begins with the first step of isolating a gene which codes for an antigenic protein of a target pathogen. This gene is then linked with a regulatory sequence that has the ability to promote high levels of gene expression. The resulting expression cassette is inserted into the plant's genome and becomes part of the genetic material present in each of the plant's cells. Plant cells with their walls removed (protoplasts), can take up foreign chromosomes or DNA directly from the environment with a very low efficiency.

The modified plant now has the ability to synthesize the foreign protein in all of the fruits or vegetables that it produces. When someone eats a fruit or vegetable from the modified plant, they are essentially eating the vaccine.

Various foreign proteins including serum albumin, human a -interferon, human erythropoetin, and murine IgG and IgA immunoglobulins have been successfully expressed in plants.

Edible plant vaccine against diarrhea, expressed in potato, and antibody against dental caries, expressed in tobacco, is already in pre-clinical human trials. Attempts are being made to express many proteins of immunotherapeutic use at high levels in plants and to use them as bio-reactors of the modern era .In the case of insulin grown in transgenic plants, it is well-established that the gastrointestinal system breaks the protein down therefore this could not currently be administered as an edible protein. However, it might be produced at significantly lower cost than insulin produced in costly bioreactors.

Overcoming the Hurdles

One of the major obstacles is developing plants that deliver sufficient quantities of a vaccine. When ingested, the edible vaccine must first maneuver through the harsh environment of the gastrointestinal tract. While oral administration of proteins has proven difficult, scientists have discovered that proteins protected by fruit and vegetable materials are able to make it through the digestive system. Even with the added protection, the stomach and intestines digest some of the proteins before they reach the immune system, so it is essential that the foods contain high concentrations of the vaccine. Getting the right amount of vaccine is also critical. Too much vaccine can lead to tolerance of disease rather than immunity and too little vaccine may not provoke an effective immune response.

Many of the first edible vaccines were synthesized in potato plants. Because raw potatoes are not very appetizing, researchers tried boiling the potatoes. But, the cooking process broke down about 50% of the proteins in the vaccine. While some proteins are more tolerant of heat, for most proteins it will be necessary to amplify the amount of protein in the engineered foods if they are to be cooked before consumption.

Tomatoes are also an excellent vehicles to deliver vaccines, because they are easy to manipulate genetically and new crops can be grown quickly. While tomatoes do not grow well in the regions in which the edible vaccines are most needed, the engineered tomatoes can be dried or made into a paste to facilitate their delivery. Genetically modified bananas are also a practical option. Children tend to like bananas and banana plants grow well in the tropical areas in which the vaccines are needed.

Transgenic plants have to grown in controlled manner by making them sterile and unable to reproduce to prevent cross pollination into our normal food supply. As an a alternative, bananas are used for genetic engineering since they do not breed and therefore their genes will not escape into the environment through seeds or pollen. Still, before they can gain FDA approval, researches must show that the edible vaccines are as effective and safe as the injectable vaccines.

But still with all these disadvantages, advancements in biotech brings us one step closer to the goal that one day edible vaccines can be used to immunize all children against the 6 most threatening diseases of measles, tetanus, diphtheria, pertussis, polio and tuberculosis. Even beyond that, there is the hope that edible vaccines can be used to conquer the spread of many other serious illnesses including yellow fever, hepatitis B virus and cholera.

Transgenic Plants

Transgenic plants or genetically modified plants are plants whose DNA is modified using genetic engineering techniques. In most cases the aim is to introduce a new trait to the plant which does not occur naturally in this species. Examples include resistance to certain pests, diseases or environmental conditions, or the production of a certain nutrient or pharmaceutical agent.

A transgenic crop plant contains a gene or genes which have been artificially inserted instead of the plant acquiring them through pollination. The inserted gene sequence, known as the transgene, may come from another unrelated plant or from a completely different species. Plants containing transgenes are often called genetically modified or GM crops.

Reasons for Making Transgenic Crop Plants:

1. The process of transgenic plant development primarily aims at assembling a combination of genes in a crop plant which will make it as useful and productive as possible. Depending on where and for what purpose the plant is grown, desirable genes may provide features such as higher yield or improved quality, pest or disease resistance, or tolerance to heat, cold and drought.

2. Combining the best genes in one plant is a long and difficult process, especially as traditional plant breeding has been limited to artificially crossing plants within the same species or with closely related species to bring different genes together.

    For example, a gene for protein in soybean could not be transferred to a completely different crop such as corn using traditional techniques. Transgenic technology enables plant breeders to bring together in one plant useful genes from a wide range of living sources, not just from within the crop species or from closely related plants.

3. This technology provides the means for identifying and isolating genes controlling specific characteristics in one kind of organism, and for moving copies of those genes into another quite different organism, which will then also have those characteristics.

4. This powerful tool enables plant breeders to do what they have always done-to generate more useful and productive crop varieties containing new combinations of genes-but it expands the possibilities beyond the limitations imposed by traditional cross-pollination and selection techniques.

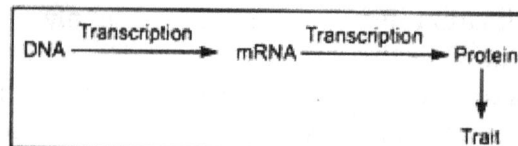

The universal concept of expression of biological traits.

Fundamentals of Transgenic Plant Development

The underlying reason that transgenic plants can be constructed is the universal presence of DNA (deoxyribonucleic acid) in the cells of all living organisms. This molecule stores the organism's genetic information and orchestrates the metabolic processes of life. Genes are discrete segments of DNA that encode the information necessary for assembly of a specific protein.

A specific protein (or an enzyme) encodes for a particular trait. In the production of a transgenic plant our primary aim is to transfer a foreign gene, encoding for some novel traits, into the genome of the plant stably.

After transferring we also need the transgene to integrate and express in the plant's cells. This process as a whole generates a new variety of plant which is new in its own kind and interests us in its massive large scale cultivation.

Steps Involved in the Production of Transgenic Plants

The fundamental steps involved in the transgenic plant production are as follows:

Step 1: Identifying, Isolation and Cloning of Genes for Agriculturally Important Traits:

The very first step in the generation of transgenic plant is to identify and isolate the novel transgene that we want to transfer into the genome of the target plant. Usually, identifying a single gene involved with a trait is not sufficient. We also have to understand how the gene is regulated, what other effects it might have on the plant, and how it interacts with other genes active in the same biochemical pathway.

Step 2: Designing Gene Construct for Insertion:

After entering the plant cell the transgene must inter-grate into the genome of the plant stably and express itself successfully so as to produce higher amount of transgenic protein which will be indirectly reflected in the trait controlled by it.

To achieve this we have to design a "gene construct" or gene-set, having all the DNA segments necessary to achieve the integration and expression of the transgene. Once a gene has been isolated and cloned (amplified in a bacterial vector), it must undergo several modifications before it can be effectively inserted into a plant.

A gene-set which will be transferred to the target plant has following segments:

A Promoter Sequence

This must be added for the gene to be correctly expressed (i.e., translated into a protein product). The promoter is the on/off switch that controls when and where in the plant the gene will be expressed. To date, most promoters in transgenic crop varieties have been "constitutive", i.e., causing gene expression throughout the life cycle of the plant in most tissues.

The most commonly used constitutive promoter is CaMV 35S, from the cauliflower mosaic virus, which generally results in a high degree of expression in plants. Other promoters are more specific and respond to cues in the plant's internal or external environment. An example of a light-inducible promoteris the promoter from the cab gene, encoding the major chlorophyll a/b binding protein.

The Transgene

Sometimes, the transgene is modified to achieve greater expression in a plant.

For example, the Bt gene for insect resistance is of bacterial origin and has a higher percentage of A-T nucleotide pairs compared to plants, which prefer G-C nucleotide pairs. In a clever modification, researchers substituted A-T nucleotides with G-C nucleotides in the Bt gene without significantly changing the amino acid sequence. The result was enhanced production of the gene product in plant cells.

Termination Sequence

This signals to the cellular machinery that the end of the gene sequence has been reached.

A Selectable Marker Gene

This is added in order to identify plant cells or tissues that have successfully integrated the transgene. This is necessary because achieving incorporation and expression of transgenes in plant cells is a rare event, occurring in just a few per cent of the targeted tissues or cells.

Selectable marker genes encode proteins that provide resistance to agents that are normally toxic to plants, such as antibiotics or herbicides. Only plant cells that have integrated the selectable marker gene will survive when grown on a medium containing the appropriate antibiotic or herbicide.

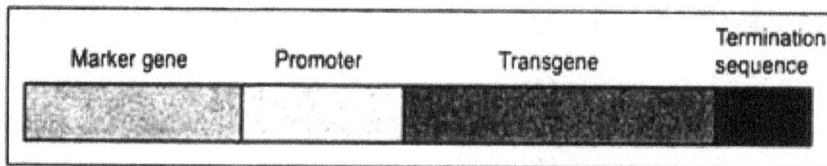

Marker gene	Promoter	Transgene	Termination sequence

The gene construct that has to be tranferred into the target plant.

The essential features of an ideal reporter gene are:

1. Lack of endogenous activity in plant cells of the concerned enzyme,

2. An efficient and easy detection, and

3. A relatively rapid degradation of the enzyme.

The commonly used selectable marker genes include those conferring resistance to the antibiotics kanamycin (nptII, encoding neomycin phosphotransferase) and hygromycin (hptIV, encoding hygromycin phosphotransferase, isolated from E. coli); and broad range herbicides glyphosate (modified versions of the enzyme EPSPS, 5-enolpyruvate shikimate-3-phosphate synthase, isolated from E. coli or Salmonella typhimurium), phosphinothricin (bar, isolated from Streptomyces hygroscopicus, codes for phophinothricin acetyltransferase), etc.

Step 3: Transforming Target Plants with the Gene Construct:

There are two ways of genetically transforming the target plant:

1. Vector mediated gene transfer, and

2. Vector less or direct gene transfer.

Step 4: Selection of The Transgenic Plant Tissue/Cells:

Following the gene insertion process, plant tissues are transferred to a selective medium containing an antibiotic or herbicide, depending on which selectable marker was used. Only plants expressing the selectable marker gene will survive and it is assumed that these plants will also possess the transgene of interest.

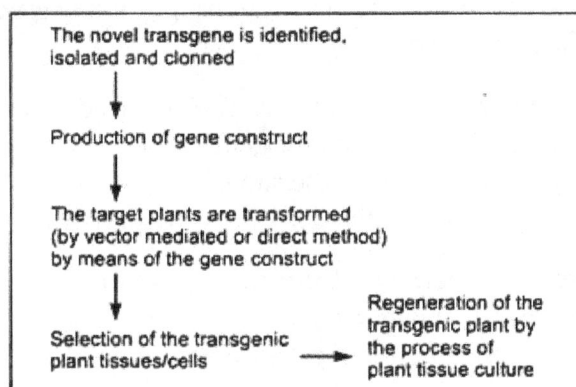

Basics involved in the production of transgenic plants

Step 5: Regeneration of the Transgenic Plants:

To obtain whole plants from transgenic tissues, they are grown under controlled environmental conditions in a series of media containing nutrients and hormones by the process of plant tissue culture.

Table: Promoters used in construction of gene construct.

Promoter	Source	Relative Activity
35S	CaMV 35S RNA gene	Constitutive, high activity: Most commonly used in dicots
35S + Adh 1 – l 1	35S promoter + first intron of maize Adh 1 gene	Enhanced promoter activity: constitutive
35S + sh 1 – l 1	35S promoter + first intron of maize shrunken 1 gene	Better than 35S + Adh 1 – l 1 in monocots: constitutive
Adh 1	Promoter of alcohol dehydrogenase gene of maize	Moderate activity in cereals: anaerobic expresion
Emu	Modified from Adh 1 promoter and its first intron	Moderate activity in cereals: anaerobic expression
Act 1 + Act – l 1	Rice actin gene + its first intron	Moderate activity constitutive
Ubi 1 + Ubi 1 – l 1 (or l 6)	Maize ubiquitin 1 gene promoter + its first (or sixth) intron	Moderate activity: constitutive High activity in cereals: constitutive
Vicilin promoter	Pea vicilin storage protein gene	Seed specific promoter

Integration of the Transgene in the Genome of the Target Plant

In general, transgenes integrate at random sites in any of the chromosomes of the genome of host cells. Usually, in a given cell, integration occurs at a single location. As a result, different cells may be expected to show integration of the transgene at different chromosomal locations.

The number of copies integrated per genome ranges from one to several hundred. In general, multiple copies are integrated when large amounts of DNA are used for transfection, while single copies are integrated with smaller amounts.

When multiple copies are integrated, they are mostly integrated at one site joined to each other head-to-tail, i.e., as a concatemer. However, in a small proportion of cases, the multiple copies are located at several sites in the same genome.

The mechanism of random integration is not known. The entire gene construct, including the vector DNA, becomes integrated. When two different gene constructs are mixed and used for transfection, they tend to be integrated together at the same site; this is known as co-transfection. The sequences flanking a gene on either side influence the expression of this gene.

Therefore, the same transgene integrated at different locations in the genome may show different levels of expression; this is known as position effect. Transgene integration frequently leads to various forms of rearrangements, e.g., duplication, deletion, etc., near the site of integration.

If these changes are large enough, the host gene located at the site of integration may become non-functional. A host gene would also become non-functional if the transgene becomes integrated within the coding region of this gene. When integration of a transgene leads to the loss of function of a host gene, it is called insertional mutagenesis; it often produces aberrant phenotypes.

Analysis of Transgene Integration

The integration of transgene into the genome is confirmed by Southern hybridization of genomic DNA extracted from the considered transgenic individuals. The DNA is digested with a suitable restriction enzyme prior to electrophoresis.

By choosing appropriate restriction enzymes for DNA digestion, not only the integration of transgene can be established beyond doubt, but information on the number of copies per cell, the orientations of tandemly arranged copies and the presence of single or multiple integration sites is also obtained from Southern hybridization. All the individuals that give positive result with Southern hybridization are regarded as confirmed transgenic.

Detection of mRNA Expression

The mRNAs produced by transgenes is most readily detected if they are with unique sequences, which have no counterparts among those produced by the host genome. A high purity RNA preparation is obtained from the appropriate tissue of transgenic individuals, and is subjected to RNA dot blot hybridization with a radioactive probe specific for the transgene.

Alternatively, the RNA preparation may be used for northern hybridization, which provides additional information on transcript size as well.

Inheritance of Transgenes

The transgenes which are stably integrated are inherited in a Mendelian fashion. They are usually dominant. Instability may occur due to point mutation, like methylation, or rearrangements of the T-DNA region. In addition, homologous recombination between copies of the transgene inserted in the same nucleus can also lead to instability of the gene.

Future Development of Transgenic Technology

New techniques for producing transgenic plants will improve the efficiency of the process and will help resolve some of the environmental and health concerns.

Among the expected changes are the following:

a. More efficient transformation, that is, a higher percentage of plant cells will successfully incorporate the transgene.

b. Better marker genes to replace the use of antibiotic resistance genes.

c. Better control of gene expression through more specific promoters, so that the inserted gene will be active only when and where needed.

d. Transfer of multi-gene DNA fragments to modify more complex traits.

Roles of Transgenic Plants as Bioreactors

Transgenic Plants

Production of Vaccine Antigens in Plants:

The aim of vaccination is to prevent infectious diseases. It can be considered as one of the most successful breakthrough of this century in the medical field. The principle of vaccination involves mimicking an infection in such a way that the specific natural defence mechanism of the host against the pathogen gets activated but the host remains free of the disease that normally results from such infection.

Followings are some vaccine antigens produced in plants:

a. Escherichia Coli Labile Toxin:

E. coli labile toxin (LT) is responsible for causing diarrhoea. It is composed of a 27 kDa A subunit and five 11.6 kDa B subunits that pentamerise. The B pentamer (LT-B) has been used as a vaccine component, as antibodies against this would block toxin activity. LT-B was expressed in transgenic potato.

b. Vibrio Cholera Toxin:

Cholera toxin (CT) is very similar to E. coli LT. The genes encoding CTA and CTB were amplified by PCR and then cloned into plant expression vectors. Expression of these genes was controlled by the CaMV strong promoter.

c. Hepatitis B Virus Surface Antigen:

Hepatitis B virus infection causes acute or chronic hepatitis, and hepatocellular carcinoma. The first vaccine developed, consists of hepatitis B surface antigen (HBsAg). This antigen is produced in large amounts in liver cells of infected individuals. Tobacco plants were genetically transformed with the gene encoding HBsAg, which was driven by CaMV strong promoter.

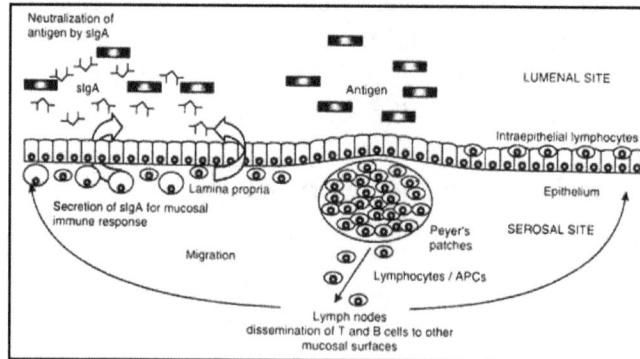

Mechanism of action of edible vaccines.

Transgenic Plants

Edible Vaccines

Vaccines that one can eat along with plant products like fruits and vegetables are called as edible vaccines. Edible vaccines are currently being developed for many human and animal diseases, including measles, cholera, foot and mouth disease and hepatitis B and C.

Mechanism of Action

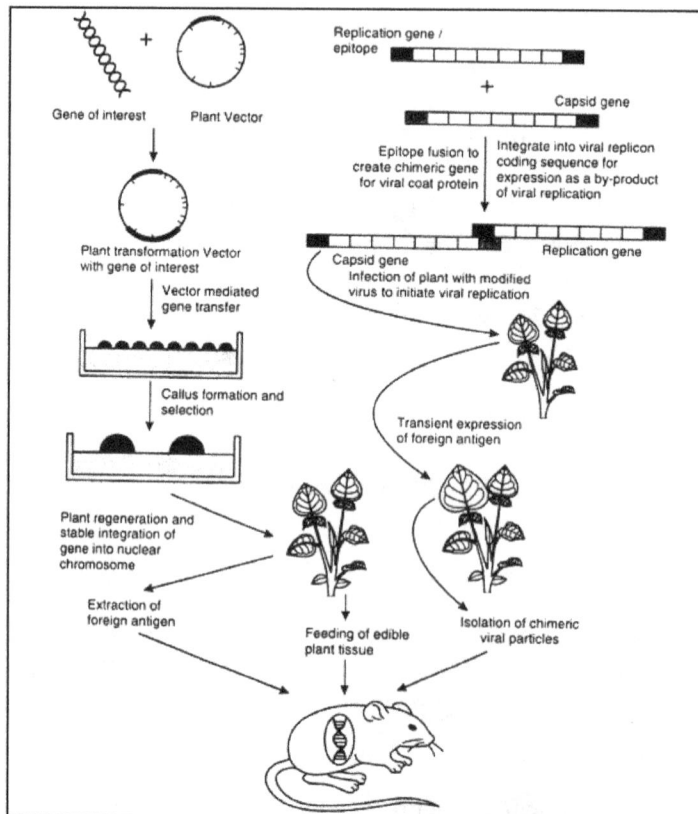

Strategies for the production of edible vaccines in plant tissues.

Edible vaccines when taken orally undergoes mastication process. Majority of the plant cells degradation occurs in the intestine near the Peyer's patches (PP), by the action of digestive or bacterial enzymes on edible vaccines. PP consists of 30-40 lymphoid nodules on the outer surface of the intestine and contains follicles from which germinal centre develops upon antigenic stimulation.

These follicles act as the sites from which antigen penetrates the intestinal epithelium, thereby accumulating antigen within organized lymphoid structure. The antigen then comes in contact with M- cells. M-cells express class II MHC molecules and antigens transported across the mucous membrane by M-cells can activate B-cells within these lymphoid follicles.

The activated B-cells leave the lymphoid follicles-and migrate to diffuse mucosal associated lymphoid tissue (MALT) where they differentiate into plasma cells that secrete the IgA class of antibodies.

These IgA antibodies are transported across the epithelial cells into secretions of the lumen where they can interact with antigens present in the lumen.

Transgenic Plants

Antibody Production in Plants (Plantibody)

Plants are used as bioreactors for the large scale production of antibodies, called as plantibodies, for the following reasons:

1. Plants can assemble heavy and light chains into complete antibodies.

2. Plants permit appropriate post-translational modification for the production of antibodies.

3. The extraction of antibody from plant cells.

4. Genetically stable seed stocks of antibody producing plants can be produced and stored indefinitely at low cost.

Production of a fully functional antibody in plants is not a very straightforward task because of the multi-subunit structure of antibody molecule. Moreover, expression of a complete antibody may not be required for many applications.

Production of Biodegradable Plastics (Bio-Plastic)

In response to the problems associated with plastic waste and its effect on the ecosystem, there has been immense focus on the development and production of biodegradable plastics. Degradable plastics can be classified either as photodegradable or biodegradable.

Photo-degradation leads to breakdown of the polymer into smaller non-degradable fragments. On the other hand, biodegradable polymer can be degraded by non-enzymatic hydrolysis or by the action of enzyme secreted by microorganisms. Among the biodegradable plastics available, there is much interest in the group of polyhydroxyalkanoates (PHAs) polymers, produced entirely from bacterial fermentation.

However, disadvantages in microbial production via fermentation include the high cost and low production. With the recent advances in molecular biology and genetic engineering allowing the possibility of using crop plants, the PHA biosynthetic genes from a bacterium (Alcaligeneseutropus) were expressed in Arabidopsis thaliana.

Table: Examples of pharmaceutical antibodies produced in transgenic plants.

Antigen	plant	Antibody	Application
Streptococcus surface antigen SAI/II	Tobacco	slgA/G (CaroRY)	Therapeutic (Topical)
Herpes Simplex virus	Soybean, rice	lgG	Therapeutic (Topical)
Non-Hodgkins lymphoma idiotypes	Tobacco, virus vector	scFv	Vaccine
Human lgG	Alfalfa	lgG	Diagnostic
CEA	Tobacco, rice wheat, pea, tomato	scFv	Therapeutic/diagnostic
CEA	Tobacco (transient and stable)	Diabody	Therapeutic/diagnostic
Respiratory syncytial virus	com	lgG	Therapeutic(inhaled)
Clostridium difficile	Com	lgG	Therapeutic (Oral)
Sperm	Com	lgG	Contraceptive(topical)
Various	Com	lgG	Therapeutic/diagnostic
Colon cancer antigen	Tobacco	lgG	Therapeutic/diagnostic

Advantages of Biodegradable Plastic

Followings are the advantages of biodegradable plastics:

1. They take less time to break down.
2. They are renewable.
3. They are good for the environment.
4. They require less energy to produce.
5. They are easier to recycle.
6. They are not toxic.
7. They reduce dependence on foreign oil.

Limitations of Biodegradable Plastics

1. Bio-plastics are designed to be composted, not recycled. The plant-based material will actually contaminate the recycling process if not separated from conventional plastics such as soda bottles and milk jugs.

2. Home composting may not be an option. Some bio-plastics cannot be broken down by the bacteria in our backyards. Polyethylene (PE) made from cane sugar is one example of this. Only bio-plastics that are fully biodegradable will break down in a home compost pile, and it could still take up to two years for certain items (e.g., forks and spoons). The rest require the high heat and humidity of an industrial composting facility.

3. Plants grown for bio-plastics have negative impacts of their own. Bio-plastics are often produced from genetically modified food crops such as corn, potatoes, and soybeans, a practice that carries a high risk of contaminating our food supply. Also, corn and soybean producers typically apply large amounts of chemical pesticides and fertilizers that pollute our air and water.

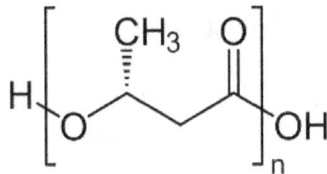

Structure of Poly-(R)-3-hydroxybutyrate (P_3HB). a polyhydroxyalkanoate.

Plantibodies

Antibodies or immunoglobulins are the vital component of adaptive immune system in mammals. These are mainly present in the body fluid and constitute an assembly of glycoproteins released by B-cells. B-cells constitute the humoral type of adaptive immune system. They are highly specific their mode of action. First they recognize their target specifically, binds to the antigen/toxin of pathogen and produce and elicit the immune response. These features permit immunoglobulins to be employed in diverse range of applications such as diagnosis, prevention and treatment. Production of defective antibody response results in increased vulnerability to pyogenic infections due to impairment in B-cell function, e.g. in hyper-IgM syndrome as a result of improper signaling between B and T cells and in X-linked agammaglobulinemia. To overcome these defects, constant region of a human immunoglobulin is modified by transgenic approach to create recombinant antibody. With this advancement, a new expectation of a reliable, inexpensive cure for diseases like, diabetes and cancer was awakening; but the major drawback was its cost ineffectiveness.

Hence, production of cost-effective and scalable platforms that is safe for therapeutics is urgently required. Transgenics is one of the most potential applications in therapeutics for the preparation of various biological substances such as antibodies, proteins and vaccines. They can be produced from plants by transforming antibody-coding genes from humans to plants as 'transgene'. Plants acts as suitable host for production of antibodies referred as "plantibodies". The term "plantibodies" describes plant based products which contained antibody and its fragments; produced by genetically engineering. Hence, plants are largely used as a host or bioreactors by exploiting their endomembrane and secretory system to generate huge amount of biological proteins of clinical importance (either full-length or smaller length fragments). This opens a new era in plant biology research. A wide variety of plants functionally expressed different types of recombinant by exploiting the same pathway for the assembly of heavy chain and light chain signal peptides followed by proper folding and assembly as in mammalian cells.

Benefits of using Plants as Host for Mammalian Antibody Production

Plantibodies functions in a similar way to mammalian antibodies. It offers numerous unique advantages over conventional methods. Firstly, plants are less prone to mammalian pathogens; this property reduces screening costs for bacterial toxins, prions and viruses. Both plants and humans have a similar endomembrane and secretory pathway system. Secondly, plants are widely distributed; their maturation time completes in a short period of time, say for example, in one season; which enables their production in a short period of time. Hence they are the cost-effective compared to their animal counterparts. Exposure to mammalian antibodies do not trigger plants immune response compared as compared to mammalian system. Production of large amount of antibodies, comparatively in short time period is the most fascinating benefit of plants as host. Comparatively, crops, for example, corn and tobacco which have higher biomass may act as good candidate in genetic engineering process; where biological products and viable proteins (stored in seed for an indefinite period) will be produced in large amounts with small decline in their catalytic activity.

Plants viz. corn, tobacco, soybean and other crops shows potential alternative for production of therapeutic proteins. Crops having maximum biomass/hectare such as alfalfa and tobacco may be a best alternative for antibodies production. Tobacco, a non-feed crop proves to be the most potential crop for large scale production of therapeutic antibodies. In transgenic tobacco, antibodies against melanoma, human papillomavirus, B-cell lymphoma, colon, testicular have been produced. These tobacco based plantibodies are still under progress for commercial use in human health.

Uses of Plantibodies in Medicine

Genetically engineered plants/transgenic plants used as host for plantibodies production represents a huge prospect for the pharmaceuticals. Among all the pharmaceutical compounds, maximum contribution is accompanied by recombinant proteins. Presently, clinical trials are going on a number of plantibodies for their therapeutic role. In therapeutics, CaroRx was the first plantibody produced from tobacco. It is anti-Streptococcus mutans secretory antibody and protects from dental caries. Another plantibody was developed in soybean against herpes simplex virus. USDA has approved CaroRx, a plantibody for humans; expressed in tobacco, against poultry virus. Plants expressing clinical proteins and polypeptides of pharmaceutical importance such as human C protein, interferons, hormones, and cytokines are currently used to provide immunization for example, oral immunization. Edible vaccines expressed in edible tissue of plants have proved to be an excellent source for expression of desirable antigens and their fragments. This is a cost effective method to provide immunization via. oral mean. Combining these, genetically engineered plants proved to be superior system for vaccination in humans.

Biopesticide

Biopesticides, a contraction of 'biological pesticides', include several types of pest management intervention: through predatory, parasitic, or chemical relationships. The term has been associated historically with [biological control] – and by implication – the manipulation of living organisms. Regulatory positions can be influenced by public perceptions, thus:

- In the EU, biopesticides have been defined as "a form of pesticide based on micro-organisms or natural products".

- The US EPA states that they "include naturally occurring substances that control pests (biochemical pesticides), microorganisms that control pests (microbial pesticides), and pesticidal substances produced by plants containing added genetic material (plant-incorporated protectants) or PIPs".

They are obtained from organisms including plants, bacteria and other microbes, fungi, nematodes, etc. They are often important components of integrated pest management (IPM) programmes, and have received much practical attention as substitutes to synthetic chemical plant protection products (PPPs).

Types

Biopesticides can be classified into these classes:

- Microbial pesticides which consist of bacteria, entomopathogenic fungi or viruses (and sometimes includes the metabolites that bacteria or fungi produce). Entomopathogenic nematodes are also often classed as microbial pesticides, even though they are multi-cellular.

- Bio-derived chemicals. Four groups are in commercial use: pyrethrum, rotenone, neem oil, and various essential oils are naturally occurring substances that control (or monitor in the case of pheromones) pests and microbial diseases.

- Plant-incorporated protectants (PIPs) have genetic material from other species incorporated into their genetic material (i.e. GM crops). Their use is controversial, especially in many European countries.

- RNAi pesticides, some of which are topical and some of which are absorbed by the crop.

Biopesticides have usually no known function in photosynthesis, growth or other basic aspects of plant physiology. Instead, they are active against biological pests. Many chemical compounds have been identified that are produced by plants to protect them from pests so they are called antifeedants. These materials are biodegradable and renewable alternatives, which can be economical for practical use. Organic farming systems embraces this approach to pest control.

RNA

RNA interference is under study for possible use as a spray-on insecticide by multiple companies, including Monsanto, Syngenta, and Bayer. Such sprays do not modify the genome of the target plant. The RNA could be modified to maintain its effectiveness as target species evolve tolerance to the original. RNA is a relatively fragile molecule that generally degrades within days or weeks of application. Monsanto estimated costs to be on the order of $5/acre.

RNAi has been used to target weeds that tolerate Monsanto's Roundup herbicide. RNAi mixed with a silicone surfactant that let the RNA molecules enter air-exchange holes in the plant's surface that disrupted the gene for tolerance, affecting it long enough to let the herbicide work. This strategy would allow the continued use of glyphosate-based herbicides, but would not per se assist a herbicide rotation strategy that relied on alternating Roundup with others.

They can be made with enough precision to kill some insect species, while not harming others. Monsanto is also developing an RNA spray to kill potato beetles One challenge is to make it linger on the plant for a week, even if it's raining. The Potato beetle has become resistant to more than 60 conventional insecticides.

Monsanto lobbied the U.S. EPA to exempt RNAi pesticide products from any specific regulations (beyond those that apply to all pesticides) and be exempted from rodent toxicity, allergenicity and residual environmental testing. In 2014 an EPA advisory group found little evidence of a risk to people from eating RNA.

However, in 2012, the Australian Safe Food Foundation alleged that the RNA trigger designed to change wheat's starch content might interfere with the gene for a human liver enzyme. Supporters countered that RNA does not appear to make it past human saliva or stomach acids. The US National Honey Bee Advisory Board told EPA that using RNAi would put natural systems at "the epitome of risk". The beekeepers cautioned that pollinators could be hurt by unintended effects and that the genomes of many insects are still unknown. Other unassessed risks include ecological (given the need for sustained presence for herbicide and other applications) and the possible for RNA drift across species boundaries.

Monsanto has invested in multiple companies for their RNA expertise, including Beeologics (for RNA that kills a parasitic mite that infests hives and for manufacturing technology) and Preceres (nanoparticle lipidoid coatings) and licensed technology from Alnylam and Tekmira. In 2012 Syngenta acquired Devgen, a European RNA partner. Startup Forrest Innovations is investigating RNAi as a solution to citrus greening disease that in 2014 caused 22 percent of oranges in Florida to fall off the trees.

Examples:

Bacillus thuringiensis, a bacterial disease of Lepidoptera, Coleoptera and Diptera, is a well-known insecticide example. The toxin from B. thuringiensis (Bt toxin) has been incorporated directly into plants through the use of genetic engineering. The use of Bt Toxin is particularly controversial. Its manufacturers claim it has little effect on other organisms, and is more environmentally friendly than synthetic pesticides.

Other microbial control agents include products based on:

- Entomopathogenic fungi (e.g. Beauveria bassiana, Isaria fumosorosea, Lecanicillium and Metarhizium spp.).

- Plant disease control agents: include Trichoderma spp. and Ampelomyces quisqualis (a hyper-parasite of grape powdery mildew); Bacillus subtilis is also used to control plant pathogens.

- Beneficial nematodes attacking insect (e.g. Steinernema feltiae) or slug (e.g. Phasmarhabditis hermaphrodita) pests.

- Entomopathogenic viruses (e.g.. Cydia pomonella granulovirus).

- Weeds and rodents have also been controlled with microbial agents.

Various naturally occurring materials, including fungal and plant extracts, have been described as biopesticides. Products in this category include:

- Insect pheromones and other semiochemicals.

- Fermentation products such as Spinosad (a macro-cyclic lactone).

- Chitosan: A plant in the presence of this product will naturally induce systemic resistance (ISR) to allow the plant to defend itself against disease, pathogens and pests.

- Biopesticides may include natural plant-derived products, which include alkaloids, terpenoids, phenolics and other secondary chemicals. Certain vegetable oils such as canola oil are known to have pesticidal properties. Products based on extracts of plants such as garlic have now been registered in the EU and elsewhere.

Applications

Biopesticides are biological or biologically-derived agents, that are usually applied in a manner similar to chemical pesticides, but achieve pest management in an environmentally friendly way. With all pest management products, but especially microbial agents, effective control requires appropriate formulation and application.

Biopesticides for use against crop diseases have already established themselves on a variety of crops. For example, biopesticides already play an important role in controlling downy mildew diseases. Their benefits include: a 0-Day Pre-Harvest Interval (see: maximum residue limit), the ability to use under moderate to severe disease pressure, and the ability to use as a tank mix or in a rotational program with other registered fungicides. Because some market studies estimate that as much as 20% of global fungicide sales are directed at downy mildew diseases, the integration of biofungicides into grape production has substantial benefits in terms of extending the useful life of other fungicides, especially those in the reduced-risk category.

A major growth area for biopesticides is in the area of seed treatments and soil amendments. Fungicidal and biofungicidal seed treatments are used to control soil borne fungal pathogens that cause seed rots, damping-off, root rot and seedling blights. They can also be used to control internal seed–borne fungal pathogens as well as fungal pathogens that are on the surface of the seed. Many biofungicidal products also show capacities to stimulate plant host defence and other physiological processes that can make treated crops more resistant to a variety of biotic and abiotic stresses.

Disadvantages

- High specificity: which may require an exact identification of the pest/pathogen and the use of multiple products to be used; although this can also be an advantage in that the biopesticide is less likely to harm species other than the target.

- Often slow speed of action (thus making them unsuitable if a pest outbreak is an immediate threat to a crop).

- Often variable efficacy due to the influences of various biotic and abiotic factors (since some biopesticides are living organisms, which bring about pest/pathogen control by multiplying within or nearby the target pest/pathogen).

- Living organisms evolve and increase their resistance to biological, chemical, physical or any other form of control. If the target population is not exterminated or rendered incapable of reproduction, the surviving population can acquire a tolerance of whatever pressures are brought to bear, resulting in an evolutionary arms race.

- Unintended consequences: Studies have found broad spectrum biopesticides have lethal and nonlethal risks for non-target native pollinators such as Melipona quadrifasciata in Brazil.

Types of Biopesticides

Biopesticides fall into three major categories:

Microbial Pesticides

Microbial biopesticides represent an important option for the management of plant diseases. Microbial pesticides contain a microorganism (bacterium, fungus, virus, protozoan or alga) as the active ingredient. Microbial pesticides can control many different kinds of pests, although each separate active ingredient is relatively specific for its target pest[s]. For example, there are fungi that control certain weeds, and other fungi that kill specific insects. They suppress pest by producing a toxin specific to the pest,causing a disease., Preventing establishment of other microorganisms through competition or Other modes of action.

The most widely known microbial pesticides are varieties of the bacterium Bacillus thuringiensis, or Bt, which can control certain insects in cabbage, potato, and other crops. Bt produces a protein that is harmful to specific insect pest. Certain other microbial pesticides act by out-competing pest organisms. Microbial pesticides need to be continuously monitored to ensure that they do not become capable of harming non-target organisms, including humans. organisms. Bt can be applied to plant foliage or incorporated into the genetic material of crops and as discovered, it is toxic to the caterpillars (larvae) of moths and butterflies. These also can be used in controlling mosquitoes and black flies. Several strains of Bt have been developed and now strains are available that control fly larvae. While some Bt's control moth larvae found on plants, other Bt's are specific for larvae of flies and mosquitoes. The target insect species are determined by whether the particular Bt produces a protein that can bind to a larval gut receptor, thereby causing the insect larvae to starve.

Plant-incorporated-protectants (PIPs)

PIPs are pesticidal substances that plants produce from genetic material that has been added to the plant. For example, scientists can take the gene for the Bt pesticidal protein, and introduce the gene into the plants own genetic material. Then the plant, instead of the Bt bacterium manufactures the substance that destroys the pest. Both the protein and its genetic material are regulated by EPA; the plant itself is not regulated.

Biochemical Pesticides

These are naturally occurring substances such as plant extracts, fatty acids or pheromones that control pests by non-toxic mechanisms. Conventional pesticides, by contrast, are synthetic materials that usually kill or inactivate the pest. Biochemical pesticides include substances that interfere with growth or mating, such as plant growth regulators, or substances that repel or attract pests, such as pheromones. Because it is sometimes difficult to determine whether a natural pesticide controls the pest by a non-toxic mode of action, EPA has established a committee to determine whether a pesticide meets the criteria for a biochemical pesticide. Biochemical pesticides include substances, such as insect sex pheromones, that interfere with mating, as well as various scented plant extracts that attract insect pests to traps. Man-made pheromones are used to disrupt insect mating by creating confusion during the search for mates, or can be used to attract male insects to traps. Pheromones are often used to detect or monitor insect populations, or in some cases, to control them.

Advantages of using Biopesticides

- Biopesticides are usually inherently less toxic than conventional pesticides.

- Biopesticides generally affect only the target pest and closely related organisms, in contrast to broad spectrum, conventional pesticides that may affect organisms as different as birds, insects and mammals.

- Biopesticides often are effective in very small quantities and often decompose quickly, resulting in lower exposures and largely avoiding the pollution problems caused by conventional pesticides.

- When used as a component of Integrated Pest Management (IPM) programs, biopesticides can greatly reduce the use of conventional pesticides, while crop yields remain high.

Bioherbicides

Weeds are a constant problem for farmers. They not only compete with crops for water, nutrients, sunlight, and space but also harbor insect and disease pests; clog irrigation and drainage systems; undermine crop quality; and deposit weed seeds into crop harvests. If left uncontrolled, weeds can reduce crop yields significantly.

Farmers fight weeds with tillage, hand weeding, synthetic herbicides, or typically a combination of all techniques. Unfortunately, tillage leaves valuable topsoil exposed to wind and water erosion, a serious long-term consequence for the environment. For this reason, more and more farmers prefer reduced or no-till methods of farming.

Similarly, many have argued that the heavy use of synthetic herbicides has led to groundwater contaminations, death of several wildlife species and has also been attributed to various human and animal illnesses.

The use of bioherbicides is another way of controlling weeds without environmental hazards posed by synthetic herbicides. Bioherbicides are made up of microorganisms (e.g. bacteria, viruses, fungi) and certain insects (e.g. parasitic wasps, painted lady butterfly) that can target very specific weeds. The microbes possess invasive genes that can attack the defense genes of the weeds, thereby killing it.

The better understanding of the genes of both microorganisms and plants has allowed scientists to isolate microbes (pathogens) whose genes match particular weeds and are effective in causing a fatal disease in those weeds. Bioherbicides deliver more of these pathogens to the fields. They are sent when the weeds are most susceptible to illness.

The genes of disease-causing pathogens are very specific. The microbe's genes give it particular techniques to overcome the unique defenses of one type of plant. They instruct the microbe to attack only the one plant species it can successfully infect. The invasion genes of the pathogen have to match the defense genes of the plant. Then the microbe knows it can successfully begin its attack on this one particular type of plant. The matching gene requirement means that a pathogen is harmless to all plants except the one weed identified by the microbe's genetic code.

This selective response makes bioherbicides very useful because they kill only certain weed plants that interfere with crop productivity without damaging the crop itself. Bioherbicides can target one weed and leave the rest of the environment unharmed.

The benefit of using bioherbicides is that it can survive in the environment long enough for the next growing season where there will be more weeds to infect. It is cheaper compared to synthetic pesticides thus could essentially reduce farming expenses if managed properly. It is not harmful to the environment compared to conventional herbicides and will not affect non-target organisms.

With the advances of genetic engineering, new generation bioherbicides are being developed that are more effective against weeds. Microorganisms are designed to effectively overcome the weed's defenses. Weeds have a waxy outer tissue coating the leaves that microorganisms have to penetrate in order to fully infect the weeds. Through biotechnology, these microorganisms will be able to produce the appropriate type and amount of enzymes to cut through the outer defenses. Streamlining of the microbe's plant host specificity will ensure that the weeds are taken out and not the crops. On the other hand, microbes can also be made to be effective against several host weeds and not only to one type of weed as this can be too expensive to produce for commercial use.

Biofertilizers

In nature, there are a number of useful soil micro organisms which can help plants to absorb nutrients. Their utility can be enhanced with human intervention by selecting efficient organisms, culturing them and adding them to soils directly or through seeds. The cultured micro organisms packed in some carrier material for easy application in the field are called bio-fertilisers. Thus, the critical input in Biofertilisers is the micro organisms.

Benefits of Biofertilizers

Bio-fertilisers are living microorganisms of bacterial, fungal and algal origin. Their mode of action differs and can be applied alone or in combination.

- Biofertilizers fix atmospheric nitrogen in the soil and root nodules of legume crops and make it available to the plant.

- They solubilise the insoluble forms of phosphates like tricalcium, iron and aluminium phosphates into available forms.

- They scavenge phosphate from soil layers.

- They produce hormones and anti metabolites which promote root growth.

- They decompose organic matter and help in mineralization in soil.

- When applied to seed or soil, biofertilizers increase the availability of nutrients and improve the yield by 10 to 25% without adversely affecting the soil and environment.

Characteristics Features of Common Biofertilizers

- Rhizobium: Rhizobium is relatively more effective and widely used biofertilizer. Rhizobium, in association wit legumes, fixes atmospheric N. The legumes and their symbiotic association with the rhizobium bacterium result in the formation of root nodules that fix atmospheric N. Successful nodulation of leguminous crop by rhizobium largely depends on the availability of a compatible stain for a particular legume. Rhizobium population in the soil is dependent on the presence of legumes crops in field. In the absence of legumes the population of rhizobium in the soil diminishes.

- Azospirillum: Azospirillum is known to have a close associative symbiosis with the higher plant system. These bacteria have association with cereals like; sorghum, maize, pearl millet, finger millet, foxtail millet and other minor millets and also fodder grasses.

- Azotobacter: It is a common soil bacterium. A. chrococcum is present widely in Indian soil. Soil organic matter is the important factor that decides the growth of this bacteria.

- Blue Green Algae (BGA): Blue green algae are referred to as rice organisms because of their abundance in the rice field. Many species belonging to the genera, Tolypothrix, Nostic, Schizothrix, Calothrix, Anoboenosois and Plectonema are abundant in tropical conditions. Most of the nitrogen fixation BGA are filamenters, consisting of chain of vegetative cell including specialized cells called heterocyst which function as a micronodule for synthesis and N fixing machinery.

Biofertilizers Recommended for Crops

- Rhizobium + Phosphotika at 200 gm each per 10 kg of seed as seed treatment are recommended for pulses such as pigeonpea, green gram, black gram, cowpea etc, groundnut and soybean.

- Azotobacter + Phosphotika at 200 gm each per 10 kg of seed as seed treatment are useful for wheat, sorghum, maize, cotton, mustard etc.

- For transplanted rice, the recommendation is to dip the roots of seedlings for 8 to 10 hours in a solution of Azospirillum + Phosphotika at 5 kg each per ha.

Application of Biofertilizers to Crops

Seed Treatment

Each packet (200g) of inoculant is mixed with 200 ml of rice gruel or jaggery solution. The seeds required for one hectre are mixed in the slurry so as to have uniform coating of the inoculants over the seeds and then shade dried for 30 minutes. The treated seeds should be used within 24 hous.

One packet of inoculant is sufficient to treat to 10 kg seeds. Rhizobium, Azospirillum, Azotobacter and Phosphobacteria are applied as seed treatment.

Seedling Root Dip

This method is used for transplanted crops. Five packets (1.0 kg) of the inoculants are required for one ha and mixed with 40 litres of water. The root portion of the seedlings is dipped in the solutions for 5 to 10 minutes and then transplanted. Azospirillum is used for seedling root dip particularly for rice.

Soil Treatment

4 kg each of the recommended biofertilizers are mixed in 200 kg of compost and kept overnight. This mixture is incorporated in the soil at the time of sowing or planting.

Use of VAM Biofertilizer

- The inoculum should be applied 2-3 cm below the soil at the time of sowing.

- The seeds are sown or cuttings planted just above the VAM inoculums so that the roots may come in contact with the inoculums and cause infection.

- Bulk inoculums of 100gm is sufficient for one meter square area.

- Seedlings raised in the polythene bags need 5-10 g of bulk inoculums for each bag.

- At the time of planting of saplings, VAM inoculums is to be applied at the rate of 20g / seedling in each spot.

- In the existing tree, inoculums of 200g is required for each tree.

Use of Blue Green Algae (BGA)

- Algal culture is applied as dried flakes at 10 kg/ha over the standing water in field rice.

- This is done two days after transplanting in loamy soils and six days after planting in clayey soils.

- The field is kept water logged for few days immediately after algal application.

- The biofertilizer is to be applied for 3-4 consecutive seasons in the same field.

Use of Azolla

- Green manure: Azolla is applied at 0.6-1.0 kg/m2 (6.25-10.0 t/ha) and incorporated before transplanting of rice.

- Dual crop: Azolla is applied at of 100 g/m2 (1.25t/ha), one to three days after transplanting of rice and allowed to multiply for 25-30 days. Azolla fronds can be incorporated into the soil at the time of first weeding.

Types of Biofertilizers

Microbial Biofertilisers

Microbial biofertilisers are biologically active (living or temporarily inert) inputs and contain one or more types of beneficial microorganisms such as bacteria, algae or fungi. Every microorganism – and hence each type of biofertiliser – has a specific capability and function. It would be relevant to mention that vermicompost is not a biofertiliser as is propagated by some, but merely an improved form of compost.

There are broadly seven types of biofertilisers:

1. Rhizobia: Rhizobia is a group of bacteria that fixes nitrogen in association with the roots of leguminous crops. Rhizobia can fix 40-120 kgs. of nitrogen per acre annually depending upon the crop, rhizobium species and environmental conditions. They help improve soil fertility, plant nutrition and plant growth and have no negative effect on soil or the environment. Every leguminous crop requires a specific rhizobium species.

2. Azotobacter: Azotobacter is also a group of nitrogen-fixing bacteria but unlike rhizobia, they do not form root nodules or associate with leguminous crops. They are free-living nitrogen fixers and can be used for all types of upland crops but cannot survive in wetland conditions. In soils of poor fertility and organic matter, azotobacter need to be regularly applied. In addition to nitrogen-fixation, they also produce beneficial growth substances and beneficial antibiotics that help control root diseases.

3. Azospirillum: Like azotobacter, azospirillum species also do not form root nodules or associate with leguminous crops. They are however not free-living and live inside plant roots where they fix nitrogen, and can be used in wetland conditions. This group of microorganisms also produce beneficial substances for plant growth, besides fixing atmospheric nitrogen. Azospirillum does well in soils with organic matter and moisture content, and requires a pH level of above 6.0.

4. Blue-green algae: Blue-green algae or cyanobacteria are free-living nitrogen-fixing photosynthetic algae that are found in wet and marshy conditions. Blue-green algae are so named for their colour but they may also be purple, brown or red. They are easily prepared on the farm but can be used only for rice cultivation when the field is flooded and do not survive in acidic soils.

5. Azolla: Azolla is a free-floating water fern that fixes nitrogen in association with a specific species of cyanobacteria. Azolla is a renewable biofertiliser and can be mass-produced on the farm like blue-green algae. It is a good source of nitrogen and on decomposition, a source of various micronutrients as well. Its ability to multiply fast means it can stifle and control weeds in (flooded) rice fields. Azolla is also used as a green manure and a high-quality feed for cattle and poultry.

6. Phosphate-solubilising microorganisms: These are a group of bacteria and fungi capable of breaking down insoluble phosphates to make them available to crops. Their importance lies in the fact that barely a third of phosphorous in the soil is actually available to the crop as the rest is insoluble. They require sufficient organic matter in the soil to be of any great benefit.

7. Mycorrhiza: Mycorrhiza is a sweeping term for a number of species of fungi which form a symbiotic association with the plant root system. Of these, the most important in agriculture is vesicular-arbuscular mycorrhiza or VAM. Plants with VAM colonies are capable of higher uptakes of soil and nutrients and water. VAM strands acts as root extensions and bring up water and nutrients from lateral and vertical distances where the plant root system does not reach.

Fungal Biofertilizer

Fungal biofertilizers comprise fungal inocula either alone or in combination, exerting direct or indirect benefits on plant growth and crop yield through different mechanisms.

Mycorrhiza is a distinct morphological structure, which form mutualistic symbiotic relationships with plant roots of more than 80 per cent of plants including many important crops and forest tree species. Plants which suffer from nutrient scarcity, especially P , N, Zn, Cu, Fe, S and B develop mycorrhiza, i.e.,the plants belong to different groups such as herbs, shrubs, trees, aquatic, xerophytes, epiphytes, hydrophytes or terrestrial ones. In recent years, use of artificially produced inoculum of two dominant types of mycorrhizal fungi has increased its significance due to its multifarous role in plant growth and yield and resistance against biotic and abiotic stresses. Ectomycorrhizal (ECM) fungi form mutualistic symbioses with many tree species. ECM fungi help the growth and development of trees because the roots colonized with ectomycorrhiza are able to absorb and accumulate nitrogen, phosphorus, potassium and calcium more rapidly and over a longer period than nonmycorrhizal roots. ECM fungi help to break down the complex minerals and organic substances in the soil and transfer nutrients to the tree. ECM fungi also appear to increase the tolerance of trees to drought, high soil temperatures, soil toxins and extremes of soil pH. ECM fungi can also protect roots of tree from biotic stresses.The most commonly widespread ectomycorrhizal product is inoculum of Pisolithus tinctorius with a wide host range and their inoculum can be produced and applied as vegetative mycelium in a peat or clay carrier. Piriformospora indica is another ECM fungus used as a biofertilizer with multifaceted traits of plant growth promotion, tolerance to both abiotic and biotic stresses and increased biomass. Endomycorrhizae from mutually symbiotic relationships between fungi and plant roots, where plant roots provide carbohydrate for the fungi and the fungi transfer nutrients and water to the plant roots. The agriculturally produced crop plants that form endomycorrhizae of the vesicular-arbuscular mycorrhiza type are now called arbuscular mycorrhizal (AM) fungi. AM fungi belong to nine genera: Acaulospora, Archaeospora, Enterophospora, Gerdemannia, Geosiphon, Gigaspora, Glomus, Paraglomus and Scutellospora. AM fungi are a widespread group and are found from the arctic to tropics and are present in most agricultural and natural ecosystems. Arbuscular mycorrhiza are prominent P mobilizers, which facilitate mobilization of soluble phosphorus from distant places in soil, where plant roots cannot reach and thus increase availability of P to plants. Since mycorrhizal fungi are more efficient in the uptake of specific nutrients like P, Ca, Zn, S, N, B and resistant against soil-borne pathogens, interest in using these fungi as biofertilizers is increasing as they play an important role in plant growth, health and productivity.

Other fungal biofertilizers, which have been used to improve plant growth by enhancing phosphorus absorption in plants are phosphate solubilizing microorganisms. The commonly widespread fungi are Penicillium, Aspergillus, Chaetomium and Trichoderma species. However applications are based on their ability to supply and mobilize plant nutrients, control plant diseases and promote plant growth and development.

Algae Biofertilizer

Algae Biofertilizer represent a large group of microorganisms which are beneficial in enhancing soil productivity. They serve the purpose by fixing atmospheric nitrogen and synthesizing plant growth promoters. Bio-fertilizers are effective replacements for chemical fertilizers and are more cost effective.

An Algae Biofertilizer is a natural, organic and renewable energy source. They help retain essential nutrients and water in the soil which is required for the proper growth of the plants. When chemical fertilizers are used they change the soil composition causing contamination and pollution.

Algae extracts have been researched for several years to be used as fertilizers. Beneficial effects of using algae have been reported in various parts. Possibility of using liquid preparations and powdered extracts has been reported to have positive effects on cereal crops including increased crop yield, improved nutrient uptake and resistance to pests.

BGA synthesizes and liberates plant growth promoting substances such as auxins and amino compounds which stimulate plant growth. They have been used particularly for rice crops.

Cyanobacteria are one of the major components of paddy fields. The agricultural importance of BGA in rice crops is related directly with their ability to fix nitrogen and other beneficial effects for plants and soil. Nitrogen content of the soil is the second major factor affecting plant growth after water. Deficiency of nitrogen content in the soil is overcome by the addition of algae biofertilizers.

Blue-green Algae

Blue-green algae are considered the simplest, living autotrophic plants, i.e. organisms capable of building up food materials from inorganic matter. They are microscopic. Blue-green algae are widely distributed in the aquatic environment. Some of them are responsible for water blooms in stagnant water. They adapt to extreme weather conditions and are found in snow and in hot springs, where the water is 85 °C.

Certain blue-green algae live intimately with other organisms in a symbiotic relationship. Some are associated with the fungi in form of lichens. The ability of blue-green algae tophotosynthesize food and fix atmospheric nitrogen accounts for their symbiotic associations and also for their presence in paddy fields.

Blue-green algae are of immense economic value as they add organic matter to the soil and increase soil fertility. Barren alkaline lands in India have been reclaimed and made productive by inducing the proper growth of certain blue-green algae.

References

- Various-applications-of-plant-biotechnology, biotechnology: yourarticlelibrary.com, Retrieved 31 July, 2019

- Edible-Vaccines-Immunization-with-Plants-or-Food: biotecharticles.com, Retrieved 28 August, 2019

- Transgenic-plants-meaning-reasons-and-fundamentals, transgenic-plants: biotechnologynotes.com, Retrieved 4 June, 2019

- Francis Borgio J, Sahayaraj K and Alper Susurluk I (eds). Microbial Insecticides: Principles and Applications, Nova Publishers, USA. 492pp. ISBN 978-1-61209-223-2

- Roles-of-transgenic-plants-as-bioreactors, transgenic-plants: biotechnologynotes.com, Retrieved 12 April, 2019

- What-are-biopesticides, ingredients-used-pesticide-products: epa.gov, Retrieved 1 January, 2019

- Bioherbicides, biotechinagriculture: isaaa.org, Retrieved 19 February , 2019

- Biofertilizers, bioinputs-for-nutrient-management, bio-inputs, agri-inputs, agriculture: vikaspedia.in, Retrieved 31 March, 2019

- Microbial-biofertilisers: satavic.org, Retrieved 15 April, 2019

- Algae-biofertilizer, algae-bioproducts: making-biodiesel-books.com, Retrieved 20 June, 2019

Permissions

Index

www.ingramcontent.com/pod-product-compliance
Lightning Source LLC
Chambersburg PA
CBHW082030190326
41458CB00010B/3322